D0845113

NONPARAMETRIC GEOSTATISTICS

NONPARAMETRIC GEOSTATISTICS

STEPHEN HENLEY

Ph.D., M.I.M.M., F.G.S.

Director, Mineral Industries Computing Ltd,
Leighton Buzzard, UK

APPLIED SCIENCE PUBLISHERS LTD, LONDON

HALSTED PRESS DIVISION
JOHN WILEY & SONS, NEW YORK & TORONTO

First published in 1981 in Great Britain by
Applied Science Publishers Ltd

British Library Cataloguing in Publication Data

Henley, Stephen
    Nonparametric geostatistics.
    1. Mines and mineral resources—Statistical
    methods
    I. Title
    622'.13'015195   TN260

ISBN 0-85334-977-0 (APPLIED SCIENCE PUBLISHERS)
ISBN 0-470-27285-6 (HALSTED)

WITH 49 ILLUSTRATIONS

Published in the U.S.A. by
Halsted Press, a Division of
John Wiley & Sons, Inc.,
New York

Photoset in Malta by Interprint Ltd
Printed in Great Britain by Galliard (Printers) Ltd, Great Yarmouth

Far better an approximate answer to the *right* question which is often vague, than an exact answer to the *wrong* question, which can always be made precise.

J. W. TUKEY, 1962
'The future of data analysis',
Ann. Math. Stats. **33**, 1–67.

Preface

The ideas in this book have been developed over the past three
or four years while I was working at the Institute of Geological
Sciences and later for Golder Associates. During that time all
of the geological modelling and resource estimation studies I
participated in had data that were non-ideal in one respect or
another (or just plain 'dirty'): the standard ways of handling the
data with kriging or with simpler parametric methods gave reasonable results, but always there were nagging doubts and some lack
of confidence because of the corners that had to be cut in generating a model. The bimodal distribution that was assumed to be
'close enough' to normal; the pattern of rich and poor zones that
was not quite a trend yet made the data very non-stationary; and
the many plotted variograms that would not fit any standard model
variogram: these all contributed to the feeling that there should
be something that statistics could say about the cases where hardly
any assumptions could be made about the properties of the parent
population.

   This book represents the current state of development of the
nonparametric geostatistical approach and is designed for use in
industry (particularly in the mining industry) by engineers who
find difficulties in using so-called standard methods; in university

research schools, where it is hoped that the methods might be chiselled to a sharp point from the present blunt instrument that they provide; and by individual scientists who, it is hoped, will be enticed away from a long standing addiction to parametric statistics.

Although the book is about statistics it is written by a non-statistician for non-statisticians; it therefore does not include rigorous mathematical proofs, but relies a good deal on intuitive reasoning. Integral calculus and matrix algebra have been kept to a minimum, and the mathematical notation has been made as simple as possible. Examples in the text are drawn mainly from geology and mining, because this is where my experience lies, but the principles are applicable to a wide variety of fields, including geophysics, meteorology, ecology, geography and oceanography: indeed, wherever parametric geostatistics has been or can be applied, so can nonparametric geostatistics.

In any new technique, there will inevitably be loose ends which remain unanswered for a while, and nonparametric geostatistics is no exception. However, its theoretical basis establishes it as a technique which can be used in a much larger set of problems than can justifiably be attacked with the mathematically tidier but more restricted kriging methods.

One thing which has been deliberately excluded from the text is computer code. Listings of programs are, of course, available, but it is expected that a reader seriously wishing to use nonparametric geostatistics will wish to have access to the latest developments rather than use an old version which is fossilised in print.

I wish to express my appreciation to all who helped me during the preparation of this book; to Dr Peter Stokes (who is now my associate in Mineral Industries Computing Limited) and to Mr Philip Green of the Institute of Geological Sciences for many long hours of discussion; to Dr Jean-Michel Rendu and Dr Alain Maréchal; to Dr Isobel Clark whose own book inspired me to start on this one, and for permission to use the iron ore example from her book; to my many colleagues in the Natural Environment Research Council, and to Miss Sheila Heseltine in particular for the final editing and typing of the manuscript. Last but most important, I acknowledge the constant encouragement I have received from my wife, without which I might never even have begun work on the book, and would certainly not have completed it.

STEPHEN HENLEY

Contents

Standard Symbols used in the Text

A	distance, area, or volume of sampling space
a	distance, area, or volume of a part of the sampling space
d, h	a Euclidean distance
F( )	distribution function (theoretical)
F*( )	distribution function (empirical)
f( )	density function
i, j	indices of observations
(i), (j)	indices of observations reordered into increasing value of *x*
K (h)	spatial covariance at separation *h*
K (0)	variance
k	dimensionality of the sampling space (1, 2 or 3)
M	median value
N	number of observations in full data set (i.e. in whole of *A*)
n	number of observations in a local subset of the data
P	a point in space at which an estimate is to be computed
P {*}	probability of event * occurring
q	a value between 0 and 1 representing a quantile

u location in Euclidean space
X variable under consideration
x a particular value (realisation) of X
$\hat{x}$ an estimate of x

α_{ij} angle at P subtended by observations i and j
$\gamma(h)$ semivariogram value at separation h
ε error in estimate; the difference between x and $\hat{x}$
σ^2 variance
τ Kendall's rank correlation coefficient
τ^* Kendall's τ modified for the purposes of nonparametric geostatistics

CHAPTER 1

Introduction

1.1 GEOSTATISTICS

One of the problems which has faced geologists and other scientists in related disciplines for many years is that of defining quantitatively the forms of surfaces and volumes which have been sampled only at discrete points. Many techniques have been devised in attempts to solve this general problem, or to solve restricted subsets of it.

Over the past thirty years, driven by demand from the mining industry for accurate estimation of ore reserves and for an indication of the reliability of those estimates, the Theory of Regionalised Variables has been developed. This was based initially on the practical methods used in the South African gold mines, notably by D. G. Krige, but has been given a substantial theoretical basis by G. Matheron and others at the Centre de Morphologie Mathématique in Fontainebleau, France, and more recently by workers in the mining industry and research organisations worldwide. The body of theory and applied methods derived from it have come to be known collectively as 'geostatistics' (though it does not by any means encompass all statistical applications in the geosciences); the practical estimation techniques that have been developed are collectively termed 'kriging' techniques, in honour of the pioneer in the field.

The essence of the Theory of Regionalised Variables is the use of information about the autocorrelations of data — to use spatial relationships which have no expression in the techniques of 'classical' statistics. In this sense, geostatistical methods are closely related to time-series analysis, many of whose methods can be viewed as one-dimensional subsets of geostatistics. Every variant of kriging is derived from the basic concept of weighted moving averages. The objective of this volume is to show that it is possible to use alternative, nonparametric estimators in the same circumstances, and relax the assumptions inherent in the use of the linear parametric geostatistical model. Such estimators will necessarily be sub-optimal *when* all the conditions for the parametric model are satisfied, but because few assumptions are made they are robust with respect to departures from the ideal case — such as skewed distributions and spatial trends: as Gibbons (1971) writes 'With so-called "dirty data", most nonparametric techniques are, relatively speaking, much more appropriate than parametric methods'.

1.2 WHY YET ANOTHER TECHNIQUE?

The purpose of this book, therefore, is to present a new approach to the same problem as is addressed by Matheron's geostatistics, that of obtaining a quantitative description of the form of a surface or volume sampled at discrete points. It may be asked quite reasonably why any new technique is required; do the many existing techniques and their variants not allow satisfactory solutions to any practical problem that might be encountered? The answer quite simply is that they do not.

Geostatistical methods developed from the ideas of Krige and Matheron all use estimators which are linear combinations of the observed data values (sometimes transformed as in lognormal or disjunctive kriging), modified by linear weightings, and the whole family of techniques thus falls within the scope of the general 'linear model' which is the basis for most of parametric statistics. Practical methods based on regionalised variable theory (rvt) have quite rightly supplanted most of the previously used techniques for many applications; rvt guarantees (subject to the assumptions that are required) that estimated values are the best *linear* unbiased estimates. More advanced variants developed in recent years to circumvent some of the more stringent requirements also provide (approximately) optimal (nearly) unbiased estimators in some restricted *non*-linear models, though at the cost of additional assumptions and the loss of assurance that the generated estimates are optimal. In all geostatistical techniques, subjective interpretation and model-fitting comprise an essential intermediate step — study of a variogram (this will be defined in Chapter 2) computed from the data and fitting of a selected 'model' variogram to be used in the subsequent kriging step — and the optimality of any final estimates is consequently not automatically guaranteed.

However, if the assumptions are satisfied, and the data allow accurate fitting of the subjective model, the kriging techniques provide an elegant solution to the problem, and one which has proven to be of great value in mining and other fields. It is unfortunate that the cases in which all conditions are satisfied tend to be those which are either of little interest or those where the answer is known already.

Difficulties arise when data do not appear to satisfy the criteria required. In some cases it is assumed that the technique is sufficiently robust to ignore the violation of the assumptions. In other cases, the more complex variants of kriging are adopted in attempts to make the data fit the theory (echoes of the epicycles used in Ptolemaic cosmology to explain away elliptical planetary orbits?). It is not infrequently found that when such a course is adopted, the costs of obtaining solutions are greatly increased, the stability of such solutions is reduced, and their optimality is at best only approximate.

The main difficulty appears to lie in the basic assumptions which are made about the statistical properties of the data. The essence of all variants of kriging is that they transform data which violate the assumptions into new variables which do not, and then operate on these in a standard way. Validity of such techniques depends entirely on how closely and how reliably they may be made to fit the parent population in the real world, from which the finite number of observations in the data set have been drawn. In classical statistics it is assumed that the distribution is usually normal (or some simple derivative of the normal); geostatistics based on rvt makes additional assumptions about the permissible types of spatial continuity relationships, and particularly that the mean and variance of the population are independent of spatial location — or at worst are described by a *simple* spatial trend.

It is very commonly found, however, that to obtain even an approximation to this state of affairs requires definition of a population too small to be of much predictive use. There is, however, a broad class of nonparametric and distribution-free statistical methods which do not require any of the more specific assumptions required by the parametric 'traditional' statistics. For spatially distributed data, for regionalised variables, no similar 'distribution-free' alternative to kriging and its variants has yet been published.

1.3 THE ROLE OF RANDOMNESS

Central to the theory or application of any statistical method is the concept of 'probability', which in itself is a numerical treatment of the notion of 'randomness'. Now, while there are some precise mathematical definitions of probability, these do not necessarily require all that is included in the philosophical or metaphysical interpretation of randomness.

When a coin is tossed in the air a large number of times, provided that it is not weighted more heavily one side than the other, it usually is found to have landed heads up on roughly half of the total number of tosses. This is an empirical observation; another empirical observation is that as the number of tosses increases, the more usual and less rough is the relationship. We now bring in the mathematical abstraction of probability to say that *in the limit* (i.e. with an infinite number of tosses) heads would show on *exactly* half of the total number of tosses, again provided that the coin is fair. The probability of 'heads' is $\frac{1}{2}$. The mathematician extends this concept by attributing this probability $\frac{1}{2}$ to refer to each individual toss: if one were to predict the outcome of each toss, one would be *equally likely* to be right as wrong. This idea matches the intuitive concept of probability, but mathematical probability, cannot in fact be deduced from any other mathematical axioms: the definition of mathematical probability is, in itself, arbitrary.

Randomness also, it seems, is not a fundamental concept. When a coin is tossed, it is certainly *very difficult* to predict which side will land uppermost, but it is not (in principle) impossible. A set of defined forces are acting on the coin at each instant. Each force could be deduced from the motion of the coin (by using high-speed photography linked to a very fast computer — beyond present day technology but not beyond the realms of plausibility), and its future course could be plotted. On a macroscopic scale such as that of a moving coin, inanimate events can generally be predicted, given enough information about preceding states. This is very much in agreement with the viewpoint of nineteenth century physicists — even of thermodynamicists, who considered atoms of gases as perfectly elastic solid objects each moving in paths determined solely by Newtonian forces.

Indeed, on the macroscopic scale, little has changed since the nineteenth century: additional forces are now known—the strong and weak nuclear forces — but these have extremely little direct influence on events in the world, certainly at scales which interest the geologist. It is only at very small scales that *quantum* effects become apparent. One consequence of quantum theory is expressed as Heinsenberg's Uncertainty Principle: when observing an electron (or similar 'elementary' particle) it is impossible in principle as well as in fact to determine both the particle's position and its momentum. The reason for this is that to observe any object we can only observe the results of some interaction between that object and some other. To read a book we use information carried by photons of light which have collided with the book. On a macroscopic scale, such interactions have negligibly small effects on the object under observation. On the scale of an elementary particle, any interaction is with another elementary particle (whose position or momentum prior to the collision were themselves unknown). The difficulty is one of observation. Whether events at such a scale are deterministic or random is a matter for philosophical discussion: in either case the cumulative effects of many such events may be considered as deterministic.

We see therefore that at very small scales, we are unable to obtain sufficient information through observation to predict events. At progressively larger scales, the number of particles increases, but as with the number of tosses of a coin, the overall proportion of landings with heads up becomes more predictable, so with increasing numbers of particles, the overall properties become more predictable. However, the larger the system, the more likely it is to contain regions of local variation (obviously reducing the predictability). It is because geological phenomena display structure at many different scales — from single crystals and mineral grains, through rock units and mineralised zones, to 'megastructures', tectonic provinces and large scale features of the Earth's crust — and also because the processes which led to the formation of rocks have left no trace but the rocks themselves (i.e. we have *location* but not *momentum* information) that randomness has been introduced into the geological vocabulary. It has been asserted by many (e.g. Mann, 1970) that randomness is fundamental. Experience has shown that statistical mathematics usually performs better at describing geology than does deterministic mathematics. These are not, however, valid arguments that chance played a major part in shaping geological events. Rather, when it appears that geological observations follow some theoretical statistical 'law', it is that insufficient observations have been made to allow definition of the complex relationships actually present, to a sufficient degree of detail.

It is, therefore, not surprising that most geological data are — in a statistical sense — 'dirty'. Either the observations follow no regular distribution law, or they bear only superficial resemblance to one of the more common distribution laws. In a few well defined cases — some sedimentary particle size distributions, for example — there may be close adherence to the normal or some other simple distribution. In this case it is a natural result of the cumulative effect of the motion of a very large number of separate grains, each on its own determined path; though all grain paths are different, there will naturally tend to be some 'average' path for each size of grain, set by the physical constraints (river boundaries, rate of flow, Stokes' Law etc.) which will condition the distribution observed. However, departures from the 'ideal' statistical distribution may be due not to inadequacies in the data but to differences of the physical system from that which generates the 'ideal' distribution. There is no necessity that the 'ideal' distribution represents the stable condition from which any departure is an aberration.

Thus, complexity of the physico-chemical systems in geology will naturally lead to complex non-ideal 'probability' distributions. The notions of 'randomness' and 'probability' are of limited and debatable usefulness when considering the 'parent' population of geological phenomena, from which samples are drawn in studying the phenomena. Is there a place for any methods based on such concepts, applied to geology?

In any geological study, the scientist's knowledge of the phenomena is necessarily incomplete; there is always a degree of uncertainty imposed by the inability to observe the whole of a population: always conclusions are drawn from samples or finite sets of observations of a limited number of variables. This uncertainty exists whether the population studied is purely random, purely deterministic, or a mixture. There is a necessity for methods which allow the scientist to minimise the uncertainty, and statistics provides a battery of such methods. It must, however, always be kept in mind that this uncertainty is associated with the sampling process and precision of the methods of measurement and *not* with the population being studied.

1.4 THE QUEST FOR STABILITY

Given a set of observations drawn from a population with unknown but possibly complex structure, the scientist's first aim is to obtain the simplest model which fits the data reasonably well. His ultimate goal is to produce a full genetic model which describes the processes that produced the population he observed. Normally, however, he must remain content with a partial description of the processes and a simple numerical model to account for deviations from his partial predictions.

In geology, there are many process models, some well understood and following precise mathematical laws (e.g. groundwater flow), others understood only qualitatively, and some lying between these extremes. It is sometimes possible to fit such models directly to the data, but more usually, a simple statistical description of the data precedes any attempt at model-fitting. For such purposes, it is essential at the preliminary stage to make the minimum of assumptions about the data — to fit the simplest possible model.

Apart from simplicity and freedom from assumptions, an essential property of such a model is *stability*. The recorded data may be noisy because of the complexity of the phenomenon; they may also be noisy because of observational error. It is necessary to use for the first fit a method which will produce a stable, conservative estimate and not, for example, inflate the importance of a single extreme value by incorporating it in averages over a large region. An optimum descriptor of the observations will fulfill two partly contradictory objectives: it will exactly fit all recorded data points (whether or not the observations include significant error, since the amount of this error is unknown at this stage); in giving this exact fit it will provide an efficient and economical interpolation between the data points; at the same time, it will provide a stable description by minimising the disturbing effect of extreme values, or outliers. To consider the choice of statistical estimators of central value, the median will provide a more stable estimate than the mean in the presence of outliers, although in 'well behaved' data sets, the mean will give a more efficient estimate.

Chapters 2, 3 and 4 will examine the cases of estimators related to the mean and to the median, for the particular class of problems in which observations are scattered regularly or irregularly in k dimensions ($k = 1$, 2 or 3). In Chapter 5, some applications of new methods related to the median are discussed, while in Chapter 6 some alternative ideas are introduced which may allow resolution of particular interpretive problems.

CHAPTER 2

Geostatistics

2.1 IDEAS AND ASSUMPTIONS

The term 'geostatistics' has been associated most frequently with the theory and methods primarily developed by Matheron and others at the Centre de Morphologie Mathématique in Fontainebleau, France. It consists essentially of a set of theoretical ideas known as 'regionalised variable theory' and a variety of practical descriptive and estimation techniques derived from them. It is impossible to present a full account of regionalised variable theory and its applications in a single chapter; by now there are a number of excellent books on the subject, including both introductory and advanced treatments of the subject. Sources which are particularly recommended are the books by Clark (1979), Rendu (1978) — both very approachable introductions; David (1977), Journel and Huijbregts (1978); and in French, Guillaume (1977). The definitive work, which laid the foundations of regionalised variable theory, is the two volume *Traité de Géostatistique Appliquée,* by Matheron (1962, 1963).

However, basic development of the practical methods preceded any of the relevant statistical theory (e.g. Watermeyer, 1919; Truscott, 1929; Krige, 1951; Sichel, 1952; de Wijs, 1951).

Matheron built on the success of these empirical methods in developing regionalised variable theory, a body of theoretical statistics in which spatial location was for the first time considered to be of importance; the concepts of autocorrelation and autocovariance were linked to a powerful new statistic, the semivariogram.

From this theoretical basis has been developed a range of methods known by the general term of 'kriging', for estimating point values or block averages from a finite set of observed values at spatial locations on regular or irregular sampling patterns.

2.1.1 Previous Methods

Spatial data analysis has been carried out using a wide variety of techniques, some with a theoretical basis, others justified purely on empirical grounds, and some with no apparent justification. Some of these methods fit the data exactly; others provide global fits of simple surfaces and therefore cannot pass through the data points exactly. Harbaugh, Doveton and Davis (1977) give a useful account of some of these methods, as does Davis (1973); only a brief survey will be given here.

2.1.1.1 The Polygonal Method

This method is very simple and in the past has been much used in the mining industry to estimate average grades within blocks. Usually the value estimated at any location is simply the value of the closest observation. It may be thought of as a weighted average with the weight on this closest point set to one, and all other weights to zero. Royle (1978) has shown this method to produce highly biased estimates, and it is now generally recognised that it is unsatisfactory even for global estimates. If used as a point value interpolation method, it gives a discrete and not a continuous estimated surface.

2.1.1.2 Moving Average or Rolling Mean

A search area — normally square, rectangular or circular — is progressively moved across the map, and the value of the centre location is estimated by the simple arithmetic mean of all points lying within this search area. Thus the weights on all points averaged are equal to $1/m$ (given m observations within the search area), and the weights on all other observations are zero. This method again gives a discrete estimated surface, but is rather better than the polygonal method. Royle (1978) shows that this method also gives biased estimates; it has, however, been extensively used in the mining industry and is still in routine use for spatial analysis of geochemical exploration data, though it is steadily being supplanted by weighted moving average and kriging methods.

2.1.1.3 Weighted Moving Average

In this method, the weights given to observations within the search area are dependent on distance of the observation from the centre location whose value is to be estimated. Weighting functions which are conventionally used include inverse distance, inverse squared distance, other inverse powers of distance ($1/d^a$ where a is not necessarily an integer), linear decreasing to zero at the edge of the search area, and negative exponential. Simple weighted averages were used by Krige in his earlier work, and are still accepted as standard in much of the mining industry. The technique is simple, cheap to compute, and can be effective, but its efficiency and unbiasedness are not optimal (Royle, 1978) and are sensitive to changes in sample spacing. All three methods mentioned so far are perturbed unduly by isolated aberrant data values.

2.1.1.4 Directional Search Algorithms

In an attempt to make the weighted average technique more stable and less perturbed by clustered data or irregular spacings, various directional controls have been introduced to limit the number of observations used in different directions away from the point to be estimated: typically quadrant or octant searching is used, with a small number (e.g. 2) of points being selected from each direction, and a requirement that data occur in at least a given number of quadrants or octants. The technique is very effective for general contouring applications and is a standard option in many commercially available packages. However, it lacks a theoretical basis, and still retains the inherent bias and inefficiency of the weighted average method. Also, like all moving averages the estimated surface cannot truly reflect the data distribution since no values can be interpolated above the maximum or below the minimum of the recorded data.

2:1.1.5 Dip Extrapolation

This is another method which is based on the idea of a weighted average. The estimate in this case is computed by averaging a set of values obtained by extrapolating surfaces defined by pairs of surrounding observations back to the location at which the estimate is required. The idea is a good one, and it allows the generation of estimates above the maximum and below the minimum of recorded data values. It is, however, very sensitive to steep gradients possibly caused by erroneous or aberrant data values. Averaging is possibly not the best way of handling the extrapolated estimates, but in any case the method is not used very much as the estimates produced tend to be less good than simple weighted average estimates.

2.1.1.6 Trend Surface Analysis

There is an extensive literature on trend surface analysis, which became used very widely during the 1960s. The method uses a simple polynomial power series or a Fourier series to provide a best fit to the observations, by choosing an optimal set of coefficients. The 'optimal' coefficients are those which produce the lowest sum of squared deviations (of data points from the estimated surface). The method works quite well with surfaces that really *are* of simple form, but it is very sensitive to data point distribution and to aberrant values.

2.1.1.7 Local Polynomial Fit

A polynomial trend surface fitted within a local search area provides an alternative to moving averages allowing estimated values to lie outside the data range. The method can work quite well with 'well behaved' data sets, but it has no theoretical basis, and it is also very sensitive to noisy data. It is sometimes difficult to make this method provide smooth interpolated surfaces. It has, however, become a standard method for many applications and is provided as an option in many commercially available contouring packages.

2.1.1.8 Linear Programming

Dougherty and Smith (1966) proposed the use of a simple trend surface method in which the sum of absolute values of deviations is minimised, rather than the sum of squares. Because of the difficulty of handling absolute values with traditional mathematical tools, they proposed that a linear programming approach be used. The method suggested had many potential advantages, and had some theoretical basis for the geophysical data with which they were concerned. Because it is a global fit method, however, it is not suitable for estimating complex surfaces, and has never been widely used.

2.1.1.9 Newton's Statistical Prediction Technique

A method was proposed by Newton (1973) which has much in common with those to be introduced in Chapter 4. The essence of the method lies in its attempt to predict the 'most probable' value at a given location. A weighted average is computed, however, with the weights optimised according to the complexity of the data on both regional and local scales. Despite the philosophical attractiveness of the method, there are inherent practical problems, not least of which is that estimated values are computed for locations which are displaced by varying amounts from the grid point locations at which estimates are required and a distorted grid is produced. This may well be satisfactory if all that is required is a contour map, but it does not provide a usable surface model.

2.1.1.10 Other Methods

There are a host of other methods which have been, and are still being, used for spatial data interpolation. However, the most important of these are the geostatistical methods, and an account of these will occupy the remainder of this chapter.

2.1.2 Regionalised Variable Theory

Although developed more recently than some of the practical methods, the theoretical background is discussed first, as the assumptions made have largely dictated the directions that have been taken in the more recent applied techniques.

Regionalised variable theory is the statistics of a particular type of variable — the regionalised variable — which differs from an ordinary scalar random variable in that as well as its usual distribution parameters, it has a defined spatial location. Two realisations of a regionalised variable which differ in spatial location display in general a non-zero correlation, as contrasted with an ordinary scalar random variable in which successive realisations are uncorrelated.

A more formal definition of a regionalised variable may be given, as follows. An ordinary random function may be defined in terms of the probability distribution law which it follows, for example, it may be normally distributed with a particular mean and variance. If this random function fits one of the standard statistical distributions, it may be completely characterised by a small number of such parameters. For many practical applications which involve observations of continuous variables, it is assumed that the observations represent particular values of a random function which is normally distributed. In an ordinary random function, however, the spatial location is not relevant — the best prediction which can be made of any observation is that which is controlled by the form of the distribution — for the normal distribution, it would be the arithmetic mean. Of course, there might well be more than one 'best' prediction, depending on the criteria used to optimise it. Thus the arithmetic mean provides the minimum squared deviation predictor for any continuous distribution; similarly the median provides the minimum absolute deviation predictor for any continuous distribution.

When the observations are scattered in space, however, it is quite likely that the values recorded at close locations will be more similar than values recorded at widely spaced locations. There will, in general, be some autocorrelation between observations. Consider the case of two observations x_1 and x_2 recorded at two locations u_1 and $u_2 = u_1 + h$, separated by distance h. If the true surface has the same mean value everywhere, then the average value of $(x_1 - x_2)$ will be zero — those observations above the mean will exactly balance those below the mean. This may be expressed as

$$E[\,x(u)\,] = m$$

$$E[\,x(u) - x(u + h)\,] = 0$$

These equations define 'stationarity' of the first order: the random variable has the same mean value m whatever the location considered.

The covariance of x_1 and x_2 may also be defined, as the expected value of the product of deviations of the two observations from the mean:

$$E[\,\{x(u) - m\}\{x(u + h) - m\}\,] = K(h)$$

where h is the distance separating u_1 and u_2, and $K(h)$ is the covariance. If it is assumed that for every h, this spatial covariance is independent of the locations of u_1 and u_2, then full second order stationarity also holds. It should be noted that the assumption of second order stationarity requires adherence also to first order stationarity.

If second order stationarity holds, then the covariance, as h tends to zero, approaches the variance of the random variable x:

$$K(0) = E[\,\{x(u) - m\}\{x(u) - m\}\,] = \mathrm{var}(x)$$

A full second order stationarity assumption is, however, rarely justifiable, and a weaker assumption may be adopted instead. The mean value m is always unknown, and may not be constant, so that the covariance and variance cannot be computed directly. An alternative statistic may be defined which does not require this mean value. Although the expected value of $x(u) - x(u + h)$ may be zero, the expected value of the square of this difference is not necessarily zero; a consistent set of assumptions may be made, weaker than second order stationarity and called by Matheron the 'intrinsic' hypothesis:

$$E[x(u) - x(u + h)] = 0$$

$$E[\{x(u) - x(u + h)\}^2] = \text{var}[x(u) - x(u + h)] = 2\gamma(h)$$

where $\gamma(h)$ is a statistic known as the *semivariogram*. The value of γ is dependent on h, the separation between the observations; at $h = 0$, γ is necessarily zero, while with increasing h, γ will tend to increase. The functional relationship between γ and h may be defined in several different ways, as will be seen later.

If second order stationarity also holds, there is a constant finite m and the covariance $K(h)$ and variance $K(0)$ both exist. The variance, covariance and semivariogram are related by the following equation:

$$\gamma(h) = K(0) - K(h)$$

since

$$\text{var}[x(u) - x(u + h)] = \text{var}[x(u)] + \text{var}[x(u + h)]$$
$$- 2 \text{cov}[x(u), x(u + h)]$$

which reduces to

$$\text{var}[x(u) - x(u + h)] = 2 \text{var}[x(u)] - 2 \text{cov}[x(u), x(u + h)]$$
$$= 2\gamma(h)$$

Parametric geostatistics as developed by Matheron requires that the intrinsic hypothesis be satisfied, although this too may be relaxed in some cases, and to a very limited degree. Thus the semivariogram is defined and is independent of the spatial location of the observations.

In common with traditional parametric statistics, geostatistical methods based on Matheron's regionalised variable theory require that the data be normally distributed — or that it be possible to transform them into a normal distribution (as is done, for example, in both lognormal and disjunctive kriging to be discussed later).

A further assumption which is implicit in the practical applications, though frequently ignored, is that the variables are additive. All linear combinations of the x values must retain the same meaning: thus it must be meaningful to compute an unweighted or weighted arithmetic average. A grade (expressed as kg/m³ for example) is an additive variable if defined on a constant support (sample size). The logarithm of a grade is not additive, since the arithmetic mean of logarithms is not the logarithm of the arithmethic mean of the grades. A topographic elevation is an additive variable, as is the thickness of a rock unit, but permeability or porosity of a rock unit are not additive, and neither is the grain size measured in phi units.

The properties of a regionalised variable may be summarised as follows: a regionalised variable is a random function with a *defined* continuous distribution at each point in space, and with a degree of spatial continuity which may be expressed by the variance of the difference in values between observations separated by distance h. For the use of practical methods which have been developed from this concept, the relationship between this spatial variance and distance h must be constant over the whole domain which is considered. Furthermore, since these practical methods use linear combinations of the variable, such combinations must be meaningful in terms of the units in which the variable is expressed.

2.2 GEOSTATISTICAL DATA ANALYSIS

2.2.1 The Semivariogram

A considerable degree of confusion has arisen over the basic terminology of geostatistics. The function γ was defined by Matheron as the semivariogram because it is the value which is used in graphic presentation and is one-half of the spatial variance $\text{var}[x(u) - x(u + h)]$; many subsequent authors have referred to γ as the variogram, and David (1977) advocates that 'for the sake of simplicity' this common usage should be adopted. Clark (1979) adheres to the term semivariogram, and this is the term which will be used here to avoid the risk of ambiguity inherent in the use of 'variogram'; however, the plotted graph may be referred to unambiguously as 'a variogram'.

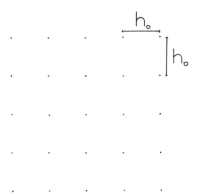

FIG. 2.1. A regular grid of observation points, with distance h_0 between rows and columns.

Clark (1979) gives a useful introduction to the computation of the semivariogram in Chapter 2 of her book, to which the reader is referred; a short account only will be presented here.

In the simplest case, a regular grid of observations is used. The interval between rows and columns is h_0 (Fig. 2.1). The value of $\gamma(h_0)$ may be estimated from such a data set by computing a sample estimate of the spatial variance $\mathrm{var}[\, x(u) - x(u + h)]$; thus if there are n_0 possible pairs of observations which are separated by a distance of exactly h_0, the semivariogram $\gamma(h_0)$ is computed by

$$\gamma(h_0) = \tfrac{1}{2}\sum[\, x(u) - x(u + h_0)]^2 / n_0$$

It is possible to compute γ for other spacings: for example, by taking alternate observations along rows and columns, $h_1 = 2h_0$ and $\gamma(h_1)$ is computed in exactly the same way using the n_1 possible pairs of observations which are separated by a distance of exactly $2h_0$. A succession of estimated values is obtained using spacings of h_0, $2h_0$, $3h_0$, ... and these may be graphed as in Fig. 2.2. It is possible to compute $\gamma(h)$ independently for different geographic directions if it is suspected that the data are anisotropic (i.e. that there is greater variation in one direction than in another), but the same computational method is used: all that

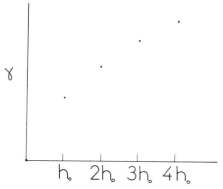

FIG. 2.2. A variogram plotted from data collected on the regular grid shown in Fig. 2.1. By definition, $\gamma(0)$ must be zero; $\gamma(h)$ tends to increase with increasing h.

changes is the number of comparisons that are used in the estimate. In the anisotropic case, h is interpreted as a vector with both magnitude and direction.

For data which are irregularly distributed in space, it is not possible in general to find sufficient pairs of observations with exactly the same spacings to generate an estimate of $\gamma(h)$ in the same way. The solution which must be adopted here is to define a tolerance or range of h values which are to be considered together; for example, all pairs of observations separated by more than 90 metres and less than 110 metres might be used in estimating the value of the semivariogram for a spacing of 100 metres. The same idea of a tolerance may also be used in computing directional γ values, by using all pairs of observations with h lying within a given distance range *and* within a given angular range to compute $\gamma(h)$.

2.2.2 Idealised Models of the Semivariogram

The semivariogram as computed from the data will tend to be rather lumpy, and the noisier the data the less regular it will appear to be. There are methods under development (Cressie and Hawkins, 1980) which aim to provide reasonably good estimates of γ in non-ideal cases. But however lumpy the graph of γ may be it can often be likened to one or other of a small number of 'ideal' curves. These curves are model semivariograms which have been defined on theoretical or empirical grounds, and the fitting of these curves to semivariograms derived from real data has been developed into the art of 'structural analysis', discussed in detail in a number of geostatistical textbooks.

The idealised curves are defined as simple mathematical functions which relate γ to h; since it is usually these ideal curves which are used in the subsequent kriging (estimation) methods, and not the semivariograms estimated directly from the data, it is worthwhile examining them in some detail, particularly with regard to additional assumptions which might be implied when using them.

2.2.2.1 The de Wijsian Model
This model is expressed as

$$(h) = A\ln(h) + B$$

Thus with increasing separation, h, the semivariogram increases without bound (Fig. 2.3). This model of the semivariogram is derived directly from de Wijs' (1951, 1953) model of the variation of ore grades in which — at any scale — a volume of ore of grade x could be divided into two parts of equal volume and grades

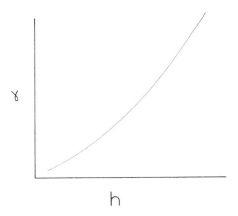

FIG. 2.3.   The general form of a semivariogram which is fitted by the de Wijsian model.

$x(1 + d)$ and $x(1 - d)$. The fraction d, if defined as a constant for all scales, leads directly to a lognormal distribution of observed grades and to the de Wijsian semivariogram model. Historically, the de Wijsian model was the first to be used by Matheron, but it usually provides a rather poor fit to the data and has been largely supplanted by the several other models now available. The de Wijsian model of ore-body grades is however still of great interest, though the simple case of constant d is not very realistic. The de Wijsian model assumes that observations and estimated values are associated with finite, non-zero volumes. It has no meaning if applied to point observations.

2.2.2.2 The Spherical Model
The spherical model is *transitive;* it consists of two separate functions, with a discontinuity (Fig. 2.4). The equations are

$$Y(h) = C\left[\frac{3h}{2a} - \frac{1h^3}{2a^3}\right] + C_0 \qquad h \le a$$

$$Y(h) = C + C_0 \qquad\qquad\qquad h > a$$

$$Y(0) = 0$$

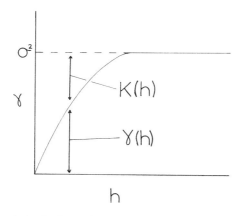

FIG. 2.4. The 'spherical' model semivariogram, showing the relationship between (classical) variance σ^2, spatial covariance $K(h)$, and the semivariogram $\gamma(h)$.

Thus for separations greater than h, there is no further increase in the semivariogram: the spatial covariance is zero, and the value of the semivariogram is thus identical to the simple variance of the complete data set, when $h > a$.

The separation a is called the *range*, $C + C_0$ is called the *sill* (numerically equal to the ordinary variance) and C_0 is a component which is known as the *nugget effect*, and represents a combination of the effects of variations on a scale below that of the sample spacings, and point errors such as errors of analysis or measurement.

The spherical model is found to fit a number of observed semivariograms to a reasonable degree of approximation (though in many cases it requires a combination of models with different values for the parameters C, C_0 and a). The spherical model itself is derived from simple geometry: the volume of intersection of two spheres of diameter a, the centres of which are h apart, is

$$V_i = V(1 - \frac{3h}{2a} + \frac{1h^3}{2a^3})$$

where V is the volume of each sphere. It is possible to set up process models to generate semivariograms which fit the spherical model, but their relationship to any real processes that generate transitive semivariograms is unexplored.

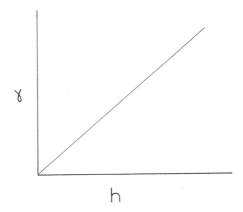

FIG. 2.5.   Linear model semivariogram.

2.2.2.3 *The Linear Model*

This is the simplest model (Fig. 2.5), and may be derived from the Brownian motion process (Feller, 1966, p.97). Its equation is simply

$$\gamma(h) = Ah + B$$

If the de Wijsian model is plotted on a log/log scale, it becomes linear; other semivariogram models also have this property: in general, the model

$$\gamma(h) = bh^{a}$$

can be transformed to

$$\ln_{\gamma}(h) = a \ln h + \ln b$$

which is identical in form to the linear model.

When such models are used to fit semivariograms derived from data, they are chosen usually on grounds of expediency, without consideration of the generating processes: indeed this almost always is the case in fitting semivariogram models to data: a combination of models is chosen to give a good fit, and little or no attention is paid to theoretical justification of the models selected.

2.2.2.4 *The Exponential Model*

Like the de Wijsian model, the semivariogram values increase with

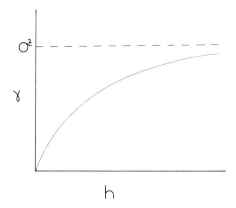

FIG. 2.6.   Exponential model semivariogram.

increasing separation h, for all h: there is no finite range. However, unlike the de Wijsian model, the semivariogram approaches asymptotically a finite limiting value (which is the ordinary global variance) (Fig. 2.6); the equation for this model is

$$\gamma(h) = C_0 + C\left[1 - \exp(-|h|/a)\right]$$

2.2.2.5 The Hole Effect Model
In this model, the semivariogram does not increase monotonically with separation, but oscillates according to the equation

$$Y(h) = C(1 - \frac{\sin ah}{ah})$$

This model is occasionally used to fit an observed semivariogram, but usually a curve such as this reflects a violation of one or more of the assumptions, such as zonal anisotropy, or a trend or drift of some sort.

2.2.3 Regularisation
It has been assumed so far that observations have been made at points. In fact, however, many types of observations are made on finite volume supports. Ore grades for example are determined from finite pieces of rock such as drill cores. The grade determined from a finite sample will be an average of the grades which would be obtained at all of the individual points within it, and this imposes a smoothing effect on the whole data set. The semivariogram values estimated for every separation h will therefore be lower than they would be if obtained from point observations. In principle it is possible to estimate the regularisation corrections required, and it is necessary to apply such corrections before using a semivariogram model for estimation. The regularisation of semivariograms is a special case of the use of the volume–variance relationship, which may be stated in general terms: the greater the volume of sample support or of a block to be estimated, the smaller the variance of samples or blocks of that volume. This relationship is examined in detail by Clark (1979, Chapter 3).

2.3 GEOSTATISTICAL ESTIMATION: KRIGING

One of the principal aims of geostatistical methods is the estimation of values at given spatial locations (points or volumes) from observations made at other locations. The kriging techniques are all related, and are refined versions of the weighted moving average techniques used by Krige. The various kriging techniques are all based on the simple linear model

$$\hat{x}_p = w_1 x_1 + w_2 x_2 + w_3 x_3 + \ldots + w_n x_n$$

where $\hat{x}_p$ is the estimator of the true value x_p at location p, and the w_i are the weights allotted to each observation, such that

$$\sum_{i=1}^{n} w_i = 1$$

2.3.1 Simple Point and Block Kriging

The difference between $\hat{x}_p$ and x_p for any set of weights w is the estimation error; for instance, if the weight on the closest observation is one, and all other weights are zero, then the value x_j of the closest observation would be used as the estimator of x_p, and the estimation error is

$$\varepsilon = x_p - x_j$$

If there is no trend (i.e. the stationarity hypothesis holds) then x_j is an unbiased estimate of x_p, and over all possible estimates like x_j the average error is zero. The variance of this difference $x_p - x_j$, however, is not zero: it is the mean of the squared difference

$$\delta_\varepsilon^2 = E\left[(x_p - x_j)^2 \right]$$

Now, since x_p is the value at location p, and x_j is the value at location j, this variance is in fact a spatial variance — it is the variogram (i.e. twice the semivariogram) at the distance h between locations p and j.

If, instead of using just one weighting of one on observation j, equal weightings were to be allocated to a number of observations, a simple arithmetic mean would be computed. However, each observation would contribute a different proportion to the total estimation variance of x_p. There are in fact an infinite number of ways in which the weighting factors can be allocated, and each will produce a different estimation variance. Among these, at least one combination of weights must produce a minimum estimation variance. It is this one which kriging seeks to find. The minimum may be determined by differentiating δ_ε^2 with respect to the weights, and setting all the resulting partial derivatives to zero; if there are n observations under consideration, there will be n unknown weights and n equations. However, the sum of the weights must be one for an unbiased estimate, so an additional equation, $\Sigma w_i = 1$, needs to be included. Because there are now $n + 1$ equations, it is desirable to obtain a well behaved system to add one more unknown — a Lagrangian multiplier. The result is a set of simultaneous linear equations which, though large, is straightforward to solve (provided that no two observations are coincident). For point kriging — that is, assuming point observations and estimating point values, the system has the form

$$w_1\gamma(h_{11}) + w_2\gamma(h_{12}) + w_3\gamma(h_{13}) + \ldots + w_n\gamma(h_{1n}) + \lambda = \gamma(h_{1p})$$
$$w_1\gamma(h_{21}) + w_2\gamma(h_{22}) + w_3\gamma(h_{23}) + \ldots + w_n\gamma(h_{2n}) + \lambda = \gamma(h_{2p})$$
$$\ldots \qquad \ldots \qquad \ldots \qquad \qquad \ldots \qquad \ldots$$
$$w_1\gamma(h_{n1}) + w_2\gamma(h_{n2}) + w_3\gamma(h_{n3}) + \ldots + w_n\gamma(h_{nn}) + \lambda = \gamma(h_{np})$$
$$w_1 \qquad + w_2 \qquad + w_3 \qquad + \ldots + w_n \qquad + 0 = 1$$

where $\gamma(h_{ij})$ is the semivariogram value for the distance h_{ij} between observations i and j; $\gamma(h_{ii}) = 0$; and $\gamma(h_{ip})$ is the semivariogram between observation i and the location p whose value is being estimated. λ is the Lagrangian multiplier. The full derivation of this system of equations is given by Matheron in several publications, by David (1977, Chapter 9), but more simply by Clark (1979, Chapters 4 and 5). Block kriging — the estimation of average block values — uses exactly the same set of equations, except that all the γ values are replaced by $\bar{\gamma}$ values. $\bar{\gamma}$ is also a semivariogram, but expresses the variance between the observation (which itself may have finite support) and the block to be estimated (of finite volume). As pointed out in Section 2.2.3, this variance will in general be less than that between points at the centres of the respective volumes, because of the averaging effect. Implicit in simple kriging of this sort, whether of point or block values, is the assumption of weak (intrinsic hypothesis) second order stationarity: there is no regional trend or drift in x values, and the semivariogram is independent of location. It must be noted that the kriging estimator is an ordinary weighted average, albeit using a particular optimum set of weights w_i as determined from the system of equations above, but it is constrained to produce a smooth surface which will not contain values which lie beyond the local upper and lower bounds of the observed data. Although it will be the best (minimum estimation variance) linear (because weighted arithmetic average) unbiased (since the weights sum to one) estimator, the kriging estimator will *not* reproduce the true values and will not give a model surface with properties which even closely resemble those of the 'true' surface.

2.3.2 When the Assumptions are False
Simple kriging as outlined above will produce the best linear

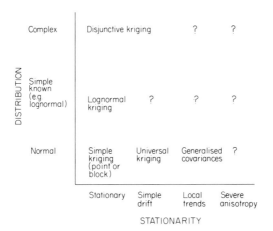

FIG. 2.7.   Kriging methods available when stationarity or distribution assumptions are not satisfied.

unbiased estimator only when all of the assumptions outlined in Section 2.1.1 are satisfied. There are many ways in which this may not be the case, however. Figure 2.7 is a schematic plot of the methods which have been developed to meet progressive breakdown of the stationarity assumption and of the normal distribution assumption. It will be seen that ways have been found to circumvent either assumption considered alone, but not together. Four of these ways will be discussed below.

2.3.3 Lognormal Kriging
It is found very often that the distribution of ore grades is not even approximately normal, but has a high positive skew and may be fitted better by a lognormal distribution. Krige (1978) considers this in detail. The ideal approach, according to Krige, is to transform the ore grades observed thus

$$y_i = \log (x_i + a)$$

where a is an arbitrary constant added as and when necessary to optimise the fit to a normal distribution. The transformed values are then used to compute semivariograms and generate the ordinary kriging estimates y_p. However, these estimates are in terms of *logarithms* of the original untransformed grades. Applying the inverse transformation to obtain an estimate of x_p will *not* produce a linear estimate of x_p from the x_i observed values: in the simplest case (a = 0) it will produce a weighted *geometric* mean, which may indeed be a good point estimator, but will not be one which minimises the estimation variance. An approximation to the desired minimum variance estimate is obtained by adjusting this value, upward, using the approximate relationship

$$\log \bar{x} \cong \overline{\log x} + \tfrac{1}{2}\sigma_x^2$$

The resulting estimator will *usually* be close to the optimum, but this is not guaranteed, and little attention has been paid to the problems which may be caused by deviation from lognormality — though Link and Koch (1975) have shown that there could be significant adverse consequences.

2.3.4 Disjunctive Kriging

Theoretically, the best possible estimator of x_p that can be obtained from a set of data x_1, x_2, ..., x_n is some function of these n values but not necessarily a linear function. This best possible estimator is known as the 'conditional expectation' of x_p given the n observations. However, in order to obtain this estimator it is necessary to know the precise distribution of the $n + 1$ variables x_p and x_1, x_2, ..., x_n. In the general case this information is not available and the conditional expectation cannot be determined. However, if it is assumed that the distribution is normal (Gaussian), and stationary, then the conditional expectation is defined and is identical with the best linear estimator — as obtained from simple linear kriging. Because of this, it would be of great use if data from any observed distribution could be transformed to a normal distribution. It is not possible to circumvent the stationarity constraint, but it is possible to obtain approximate transformations to convert an arbitrary continuous distribution to a normal distribution. A simple example of this is found in lognormal kriging; in this an ordinary logarithmic transformation is used.

In disjunctive kriging, a 'best-fit' approximate transformation is used, consisting of a set of Hermite polynomial functions (Journel and Huijbregts, 1978, pp. 573–4). It is possible by using such transformations to convert the data into a form which will approximate to univariate normal distributions for the x_i values and bivariate normal distributions for every pair of values x_i, x_j, though not to achieve full multivariate normality of x_p, x_1, x_2, ..., x_n. It is necessary therefore to restrict the class of possible functions of x_1, x_2, ..., x_n to those which are linear combinations of univariate functions of x_i:

$$\hat{x}_p = g_1(x_1) + g_2(x_2) + \ldots + g_n(x_n)$$

where the functions g_i are measurable functions which may be estimated by a straightforward system of kriging equations if it can be assumed that the data have been transformed to a normal distribution.

Disjunctive kriging provides, in principle, a better estimate than is available from ordinary linear kriging, but less good than the conditional expectation (which is the best possible estimate). In practice it can do this given stationarity — as for linear kriging — plus sufficient knowledge of the distributions of the observed values x_i to be able to obtain good approximate transformations to univariate and bivariate normal distributions. The assumptions required in practice, therefore, are more restrictive than are those for linear kriging. Moreover, there is no indication of any ways in which the stationarity constraint might be relaxed, and there is a greatly increased computational cost in disjunctive as compared with linear kriging.

It should also be noted that the estimator obtained is a *non-linear* estimator, and many of the 'nice' properties derived from the linearity of the simple linear kriging method are discarded in disjunctive kriging, as in lognormal kriging.

2.3.5 Universal Kriging
So-called 'universal kriging' is not universal either in theory or in application. However, it represents the first significant attempt to avoid the onerous requirements even of 'intrinsic hypothesis' stationarity, which are rarely adhered to in nature. The alternative assumption which is made is that the mean value, instead of being constant, follows a regular trend or drift over the area of interest, and that this trend can be expressed as a simple polynomial function of the form

$$m = a_0 + a_1 u + a_2 v + a_3 u^2 + a_4 uv + a_5 v^2 + \ldots$$

This is a function which will be familiar to those who have used polynomial trend surface analysis. Indeed, the system of equations used in universal kriging is a combination of the linear kriging system given in Section 2.3.1 and the system used for polynomial trend surface analysis. The equations are simply expressed, but there is a big problem. This is the definition of the semivariogram values to be used. The kriging equations include, for UK (universal kriging), n unknowns for the weights on the observations, one unknown for the Lagrange multiplier, and k unknowns for the coefficients on the polynomial terms: these are computed separately for each x_p to be estimated (and thus the surface will *not* be closely related to the polynomial trend surface). However, if these coefficients are unknown before solution of the equations, they cannot be used to compensate for drift in computation of the semivariogram, and every $x(u) - x(u + h)$ term will contain an unknown contribution from the deterministic drift. This can, and does, grossly distort the semivariogram as plotted. As David (1977 p. 273) notes, no direct solution to this problem is yet known, and it must be approached by a trial-and-error procedure in which an assumed semivariogram form is used, and an attempt is made to estimate a variogram of residuals from an approximated drift within a subjectively chosen neighbourhood size or range. When a self-consistent solution is obtained, it is assumed correct and used in the UK estimation.

2.3.6 Generalised Covariance

So far, all kriging methods described (apart from UK) have assumed either stationarity of the mean, or stationarity of successive differences; these two hypotheses may be expressed as

$$E[x(u)] = m = \text{constant (stationarity of the mean)}$$

and

$$E[x(u) - x(u + h)] = \frac{\delta m}{\delta u}$$
$$= \text{constant (stationarity of differences)}$$

The first of these equations is assumed in first or second order stationarity, the second (but not the first) in the intrinsic hypothesis. However, there is no necessity to restrict consideration to stationarity of these zero or first order differences: it is possible to define second, third, ... nth order finite differences which may be assumed stationary. It should be noted that stationarity of the differences of any order necessarily entails stationarity of the differences of all higher orders (since these differences are all zero). The strongest forms of stationarity, therefore, are those in which the lowest order differences are stationary. The intrinsic hypothesis is thus a weaker form of stationarity than ordinary first order stationarity. Similarly, if second order differences are assumed stationary (but not zero or first order), the assumption made is weaker still, allowing greater freedom in the forms of drift allowed in the estimated surface.

These differences of any order were termed by Matheron (1973) and Delfiner (1976) 'generalised increments'. Consider the case of regularly spaced observations on a line. A second order increment has the same form as the second order finite difference used to approximate $\partial^2 x / \partial u^2$ (in fluid flow modelling for example):

$$x(u + h) - 2x(u) + x(u - h)$$

In general, it requires $k + 1$ terms containing different observed x values to compute a kth order finite difference.

The concept of generalised increments leads directly to that of generalised *covariances:* the variance of increments of order k is expressed by a function $K_k(h)$, the generalised covariance: for $k = 0$, with simple increments, the generalised covariance is equal to the semivariogram. The kriging error itself may be considered as an increment of order k, thus the standard kriging equations may be entirely rewritten in terms of generalised covariances of any order k. The result of doing this would be to relax the stationarity requirements to an appropriate degree. The cost, however, is measured in complexity of the theory, in complexity of the computations required to obtain the generalised covariances (and thus increased computational costs) and in difficulty of subjective interpretation of intermediate results: in implementations of generalised covariance methods in packages such as BLUE-PACK, provision is made for fitting polynomial models to generalised covariances, but no models are available which match the theoretical elegance of those available for the semivariogram.

2.4 PROBLEMS WITH GEOSTATISTICS

The geostatistical methods described in this chapter are all *parametric* and are derived from regionalised variable theory. Generically they have great advantages by comparison with pre-existing techniques for analysis of spatial data; they have a sound theoretical basis, they allow some estimation of the *quality* of estimates produced, and they have some claim to statistical properties such as unbiasedness, linearity, and minimum variance. However, they require some quite stringent assumptions to be made; assumptions which even their advocates would readily admit are rarely met in nature.

2.4.1 Non-Stationarity

Fundamental regionalised variable theory requires that at least the 'intrinsic hypothesis' form of stationarity is true: it is accepted that there are local variations in the mean, but it is nevertheless required that the spatial variance — the semivariogram — is stationary over the whole region of interest. Now, as most authors accept, real data sets rarely even approach stationarity. Universal kriging and the generalised covariances method extend the capabilities to allow for non-stationary data, but even in these methods, the types and amounts of non-stationarity are restricted to a few idealised situations. Most authors on geostatistics suggest that departures from stationarity are not of practical significance since *local* stationarity may often be assumed; there is, however, no general proof of this, and no statistical test to determine whether such an assumption is warranted; the fact remains that regionalised variable theory itself is not valid under conditions where its defined form of stationarity does not exist.

2.4.2 Non-Normality of Distribution

Although normality of distribution is one of the assumptions of regionalised variable theory, it has always been so obvious that most data sets are not normally distributed, that a family of techniques has been developed in an attempt to avoid this constraint. These methods include lognormal kriging and disjunctive kriging. The common aim of such methods is to transform the data to a normal distribution before kriging with standard equations. The basic objection to such a procedure is that the variable to be estimated is then a non-linear — and perhaps quite complicated — function of the original data. Use of the standard kriging equations on transformed data will minimise some function other than a simple variance, and the resultant estimate — in data units — will in general be both sub-optimal (because it is not the minimum variance estimator) and possibly biased (because the *un*biased estimate is of a transformed value). Instead of the best linear unbiased estimator, these methods produce sub-optimal non-linear biased estimators — albeit quite often better than the best linear, but only because the class of possible non-linear estimators is so much larger than that of linear estimators.

2.4.3 Subjectivity

The cornerstone of parametric geostatistical methods is the semi-variogram. This may be computed from the data set under investigation or from some other data set which has a postulated relationship with it. Whatever data set is used, the semivariogram will quite often depart significantly from *all* of the theoretical models, and skillful interpretation is required to fit one or more models to the empirical curve (or to recognise breakdown of assumptions such as stationarity, which will have direct distorting effects on the semivariogram). If the semivariogram is computed from one data set and used for kriging another, care must be taken to ensure that the two lie within a single homogeneous region. On the other hand, if the semivariogram is generated from the data set which is to be kriged, any peculiarities in the data which produce distortions may be magnified when kriging the same data. Resolution of such problems requires subjective interaction with the data, and the quality of results will depend largely on the skill of the interpreter.

A second, related, subjective influence is that of choice of technique. If some departure from stationarity has been diagnosed from the semivariogram, it might be considered best to use universal kriging or generalised covariances; on the other hand, if it is known that the data follow some complex non-normal distribution, it would probably be deemed appropriate to use disjunctive kriging. What, then, is the rational choice of method when both situations occur simultaneously? (This situation is represented by the upper right corner of Fig 2.7.) Not only do methods not yet exist to meet such a situation, but if they were devised, they would inevitably be of fearsome computational complexity, to combine the properties of both generalised covariances and disjunctive kriging.

CHAPTER 3

The Nonparametric Approach

3.1 CONCEPTS AND ASSUMPTIONS

Both the traditional and the more recent parametric techniques in statistics require, as was emphasised in Chapter 2, that a number of assumptions be satisfied. In practice it is true that most of these techniques generate plausible, and even acceptable, solutions even when the assumptions are not wholly satisfied. However, in such cases the claims of optimality cannot be upheld with any conviction, and insufficient evidence is available to justify any great faith in the robustness particularly of the newer geostatistical methods.

In geology and other natural sciences, data are commonly obtained by measurement or observation of phenomena which are the results, in general, of complex processes. Although one may enter into a long philosophical debate over determinism versus randomness in natural events, it is generally agreed that, on a macroscopic scale, processes that tend to operate deterministically are more evident, although on the microscopic scale stochastic laws often appear to prevail. In a sandy beach, for instance, the macroscopic structure is highly ordered and is controlled by tides and currents. In contrast, the sizes of sand grains in a ten gramme sample of the same beach will follow some frequency distribution which is defined by a statistical response to the same processes: small-scale turbulence obscures the larger scale ordering. However, there is no reason, in general, for the size frequency distribution of sand grains to match any 'ideal' statistical distribution defined on the basis of mathematical convenience; indeed it is most unlikely that geological data will follow any purely statistical law. There are many cases in which data appear to fit normal, or lognormal, or gamma distributions; there are also a great many published data sets which fit no simple statistical distribution. Even when there is an apparently good fit, the choice of statistical distribution is made subjectively *after* examination of the data; rarely is there any justification for preferring one particular ideal distribution. Thus, quite often the most that can be said of the data distribution is that it is continuous, and even this statement must be qualified by stating the bounds of continuity (e.g. the proportions of chemical elements or of minerals in rock samples are continuous in the range zero to one).

It will be seen in Chapter 4 that the assumptions of spatial stationarity required by kriging methods are, similarly, made for mathematical convenience and without any assurance of validity or robustness.

Classical statistical methods such as the *t*-test and the *F*-test, and methods based on regionalised variable theory, all require such restrictive assumptions to hold (i.e. normality of the distribution, or that plus stationarity). If the data do not allow these assumptions to be made, it is unsafe to use these statistical methods. Are there any methods that can be used irrespective of the data distribution? The answer is that there is a class of techniques which may be applied and which require the minimum of assumptions: usually continuity of the distribution is sufficient (though some methods do require additional assumptions such as symmetry of the distribution). Appropriately they are called 'distribution-free' statistics. Many distribution-free methods — by their nature — are not concerned with distribution parameters, and are therefore appropriately termed 'nonparametric' statistics. The two classes, of distribution-free and nonparametric statistics, overlap to such a large extent that the terms are often used interchangeably. Gibbons (1971) makes a useful distinction between them: 'tests of hypotheses which are not statements about parameter values have no counterpart in parametric statistics, and thus here nonparametric statistics provide techniques for solving new kinds of problems. On the other hand, a distribution-free test simply relates to a different approach to solving standard statistical problems'. The methods to be described in this book are both nonparametric *and* distribution-free, and the term nonparametric is used throughout.

3.1.1 Definitions

In entering the uncharted waters of nonparametric statistics, a number of factors must be considered which are taken for granted in using classical parametric techniques. These include bias in a test or estimator, consistency of a test, efficiency (or power), as well as a much more fundamental factor — the type of measurement scale used.

3.1.1.1 Measurement Scales

There are three different types of measurement scale which are commonly used. First, the *nominal* scale records only qualitative information (e.g. rock names); in its simplest form only yes/no data are recorded. Second, the *ordinal* measurement scale is used when the possible values are placed in a numerical sequence, but where no precise statement is made about the significance of intervals between successive values: examples are Mohs' scale of hardness of minerals, the sequence of ammonite zones in the Jurassic, and the numbers of zircon grains in a set of equal sized samples of sand. Third, the *interval* scale consists of real-valued measurements on a continuous scale; in theory the measurements may take any real value within the range of the scale, but in practice this is of course not possible and approximations to the 'true' value are imposed by the degree of accuracy available on the measuring instrument. A special case of the interval scale is the *ratio* scale, in which there is an absolute zero value. Thus spatial distance is measured in an ordinary interval scale, but temperature or mass are measured in ratio scales, as are the proportions of constituents in a rock sample.

The scales may be distinguished according to the mathematical operations that are meaningful among measurements: thus addition or subtraction are valid operations on measurements in the interval scale, but are quite meaningless for nominal or ordinal measurements. Similarly, the terms 'greater than' and 'less than' have meaning when applied to measurements in the interval and ordinal scales, but not for the nominal scale.

3.1.1.2 Bias
It is commonly desirable that an estimator or test statistic, *on the average*, be close to the true value being estimated. An unbiased statistic is one whose expected value over all possible samples from a population is equal to the true population value. This is not a requirement in every case — some tests are designed to be conservative or pessimistic, with a built-in bias in a known direction.

3.1.1.3 Efficiency
For nonparametric methods which have direct counterparts in parametric statistics, there will naturally be a difference in performance. This is because the parametric methods use the full numeric information available, while nonparametric methods usually use only a part of that information — like the ordering of measurement values. Intuitively one would expect the nonparametric methods to be less efficient than corresponding parametric methods (at least when the conditions for validity of the parametric methods are satisfied). There are several objective ways to compare the efficiencies of two different unbiased estimators of a parameter. For example, the ratio of their variances may be used or the limiting value of this ratio as the sample size increases to infinity.

The *power* of a test is the probability that the null hypothesis will be rejected when it is false: the *power efficiency* of a test is the ration n_a/n_b where n_a is the number of observations required for the test to have the same power as the parametric test has with n_b observations.

A measure of efficiency of nonparametric tests which has become more or less standard is the *asymptotic relative efficiency* or ARE, defined as the limiting value of this ratio while simultaneously n_b tends to infinity and the alternative hypothesis H_1 approaches the null hypothesis H_0. As Gibbons (1971) points out, however, the ARE is a large-sample estimator and is meaningful only close to H_0; its acceptance by the statistical community is largely by default, there being no more universal measure of efficiency. That said, it is usually a reasonable guide to the performance of a nonparametric method in real situations.

3.1.1.4 Consistency

A statistical test is consistent when the power of the test approaches one as the sample size approaches infinity. This is obviously a desirable property of a test, and is automatically provided by the standard parametric tests.

3.1.2 Distribution and Density Functions

The matter of distribution functions has already been touched upon in Chapter 2, and it is assumed that the reader is familiar with the 'standard' classical distribution functions and the idea of continuous distributions.

Distribution-free techniques, such as to be described in this and the next chapter, need (by definition) no detailed assumptions or knowledge of the form of distributions, beyond the stipulation that they be continuous within a defined range. Most of the techniques achieve this by mapping the data into a simple uniform distribution and operating on the new uniformly distributed variables. How this is done may be demonstrated graphically. A sample distribution may be plotted as in Fig. 3.1 by ordering observations into increasing value and plotting rank (that is, position $r_{(i)}$ in the reordered array) against the value, $x_{(i)}$. Each point plotted gives an estimate of the value below which a given proportion $r_{(i)}/(n + 1)$ of the population occurs. As can be seen in Fig. 3.1(a), (b) and (c), the effect of ranking on observed distributions, (or on samples drawn from theoretical distributions), is to map the actual continuous distributions into a uniform distribution of ranks.

Density functions are of less direct application in nonparametric

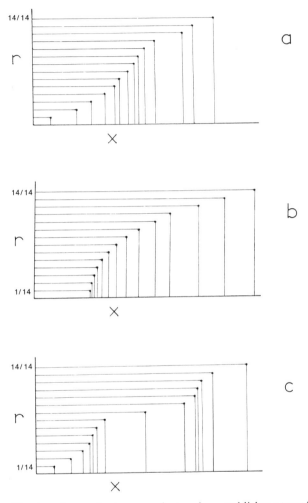

FIG. 3.1.   Plotting observed values x against ranks r establishes a transformation from any distribution (e.g. (a) symmetric unimodal, (b) asymmetric unimodal, (c) bimodal) into a uniform distribution.

statistics, and are in any case more difficult to estimate from samples. Given a moderately large sample it is possible to group observations into classes with different value ranges: if the numbers in each are plotted, the result is a histogram, which is a discrete representation of the sample estimate of the density function.

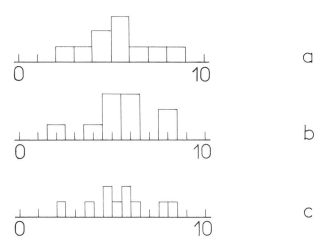

FIG. 3.2.   With small samples, the histogram is unstable and is sensitive to choice
of class limits. (a), (b), (c): different combinations of class limits and intervals,
using this set of data — 5·1, 4·8, 6·3, 5·8, 8·1, 4·9, 3·9, 2·2, 5·6, 7·8.

Unfortunately, however, the histogram is unstable for small samples as its appearance is very sensitive to choice of class limits and intervals (Fig. 3.2).

An alternative way to estimate a density function is to plot the gradient of the distribution function (since the density function is the first derivative of the distribution). This is not altogether satisfactory, however, as it is unduly sensitive to minor fluctuations in observed values, and infinite gradients are produced whenever tied values occur. It is possible to compute the gradient of a smoothed distribution curve, but since the main application of the sample density function is in preliminary visual examination of the data set, such an approach may not be worthwhile. The histogram has at least the advantage of a direct relationship with the raw data.

3.2 ESTIMATORS AND TESTS

The aims of statistics may be of two main types:

(i) to deduce from a sample some of the properties of a population, for example, estimating parameters, and

(ii) to test hypotheses about properties of the population.

Separate statistical techniques have been developed to meet each
of these aims and may correspondingly be classified into:

(i) exploratory, and

(ii) confirmatory statistics,

though a few techniques may be used in both ways. There are certain desirable properties which are shared by well designed estimators (statistics satisfying objective (i)) and tests (statistics satisfying objective (ii)); because of these shared qualities, the mathematics of the two classes of statistics tend to be very similar. The
desirable properties include:

(a) *Consistency*. As the sample size is increased, the result
obtained from the statistical procedure should converge to
a single value: the error in an estimate should tend to zero,
while the power of a test should approach one.

(b) *Unbiasedness*. Over all possible samples in the sample space
considered, the expected value of the estimator or test statistic should be the same as the corresponding parameter
derived from the complete population. The simplest case
of an unbiased statistic, for example, is the arithmetic mean
$\bar{X}$ where $E(\bar{X}) = \mu$.

It should be noted, however, as mentioned earlier, that
although unbiasedness is generally a desirable property,
many valuable statistics do in fact have an intentional bias.

(c) *Sufficiency*. The estimating or testing function (or more
generally the procedure) should be independent of the value
of the parameter being estimated or tested. Most classical
statistical methods and most standard nonparametric methods meet this criterion. Some of the adaptive techniques
to be discussed in this chapter may, at first sight, appear
to violate this criterion, but in fact if one takes the broader
interpretation (procedure rather than function), then sufficiency encompasses these also.

3.3 ESTIMATORS

There are two population parameters which require to be estimated more than any other: the location of the centre of a distribution (leaving undefined for now the term 'centre'), and the amount of spread or dispersion about this centre. The classical parametric statistics which provide these estimates are the arithmetic mean, $\bar{X}$, defined as

$$\sum_{i=1}^{n} X_i/n,$$

and the sample variance

$$\delta^2 = \sum_{i=1}^{n} (X_i - \bar{X})^2/n$$

The standard deviation, δ, is also commonly used as a measure of dispersion; it is simply the positive square root of the variance.

Other population parameters of which estimates are often required are skewness and kurtosis, parameters describing the shape of the distribution (skewness is a measure of the 'lopsidedness' and kurtosis describes the 'flatness' or 'peakedness'); again there are standard parametric estimates of these.

There is a plethora of robust and distribution-free estimators for the central location and for dispersion; some of the more useful of these are mentioned below. First, though, it will be necessary to look at some nonparametric statistics for which there is no parametric equivalent.

3.3.1 Order Statistics

Most distribution-free estimators are derived in some way from order statistics. A vector of sample data $x_1, x_2, \ldots, x_n$ may be reordered into strictly ascending value of x, to give an ordered vector $x_{(1)}, x_{(2)}, \ldots, x_{(i)}, \ldots, x_{(n)}$ such that for every i, $x_{(i)} \leq x_{(i+1)}$. ($x_{(0)}$ is taken to be $-\infty$, and $x_{(n+1)}$ is taken to be $+\infty$). Thus the smallest finite element of x is $x_{(1)}$, the second smallest is $x_{(2)}$, the ith smallest is $x_{(i)}$, and the largest is $x_{(n)}$. The operation of this ordering transformation is illustrated in Fig. 3.3. In an ideal continuous

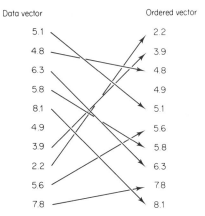

FIG. 3.3 The ordering transformation applied to the vector of data used in Fig. 3.2.

distribution every $x_{(i)}$ will be distinct, though in practice because of limits of precision of measuring instruments, ties may occur.

The n ordered values partition the population into $n + 1$ parts (every possible interval $\overline{x_{(i)}, x_{(i+1)}}$), and it can be shown that these are, on the average, of equal size: each interval contains the same proportion of the population.

The indices (i) of the order statistics are automatically distributed uniformly in the interval $\overline{0, n+1}$, and replacement of $x_{(i)}$ by i is effectively a transformation from the observed distribution into a uniform distribution. (In some nonparametric methods a second transformation is required, replacing i by $Z(i/(n+1))$ where $Z(p)$ is the standard normal variate corresponding to the cumulative proportion p of the distribution.)

The terms 'quantile' and 'percentile' refer to order statistics at any point between and including the minimum and maximum data values. The pth quantile or the $100p$th percentile of a population is a value which lies above a proportion p of the distribution and below the remaining $1 - p$. As can be seen in Fig. 3.4, is not necessarily a unique value, although for continuous distributions it generally will be. Because of the way in which order statistics partition a distribution, they may be used as unbiased estimators of quantiles: the ith order statistic $x_{(i)}$ is an unbiased estimate of the

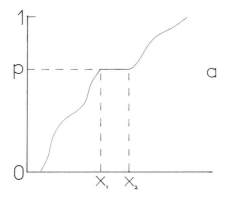

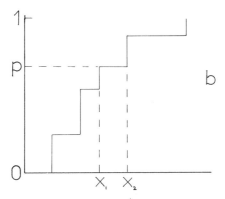

FIG. 3.4. In either (a) a continuous distribution with gaps, or (b) a discrete distribution, a quantile p may take any value between x_1 and x_2.

$i/(n + 1)$th quantile, since on average, i out of $n + 1$ equal partitions of the population will lie below $x_{(i)}$.

Order statistics have formed the basis of a large variety of empirical estimators, largely because of the computational simplicity and convenience. The sample minimum and maximum are of course $x_{(1)}$ and $x_{(n)}$ respectively. The range is $x_{(n)} - x_{(1)}$ and is the simplest (though not very good) measure of dispersion. The sample median — estimating the 0·5 quantile — is the central value $x_{((n + 1)/2)}$ if n is odd, or is any value between the two central values $x_{((n + 2)/2)}$ and $x_{(n/2)}$ if n is even; conventionally in this case the median is interpolated midway between these two values. The median will be discussed in more detail in a later section. Apart from the median, some other quantiles have been given special names. There are three quartiles, dividing the population into four equal parts; similarly the quintiles divide the population into five, and the deciles divide it into ten.

In practice, of course, the sample is unlikely to consist of an exact multiple of four, five, ten, or a hundred observations, and such quantiles are usually estimated either by taking the closest order statistic or by linear interpolation between adjacent order statistics as is done for the median.

3.3.2 Rank

The rank of an observation $x_{(i)}$ is simply its position i in the order statistic vector. Thus the smallest data value has a rank of one, the highest has a rank of n. The rth smallest data value has a rank of r. The only exception to this rule is encountered when tied observations occur. There are a number of ways of dealing with tied observations (for example assigning ranks arbitrarily, elimination of tied values from the statistical investigation, assigning to all tied observations the mean of the ranks they would have had if they had been distinct) and the purposes of these are discussed by Gibbons (1971).

3.3.3 Rank-Order Statistics

Rank-order statistics comprise all statistical functions which are derived from a set of rank values: a very large number of nonparametric methods employ functions of this type. Such methods are necessarily distribution-free, because the ranks themselves are uniformly distributed irrespective of the distribution of the sample or of the population. The actual values of observations are used only once, to establish the ordering transformation, and are disregarded thereafter. Gibbons (1971, Chapter 5) discusses the theory of rank-order statistics.

3.3.4 Estimates of Central Location

In parametric statistics there is no difficulty in defining the central location: it is the position of the peak of a normal distribution, which coincides with the median; it is estimated most efficiently by the arithmetic mean of a random sample. In fact, in any symmetric unimodal distribution, the mean, median, and mode coincide, although the mean is not the most efficient estimator for all such distributions (Hogg, 1974); for the sharp-peaked double exponential distribution, for example, the median is the maximum likelihood estimator of the centre; for the distributions close to the uniform distribution the mid-range $(x_{(1)} + x_{(n)})/2$ is the maximum likelihood estimator.

The skewed distribution — even a simple theoretical one like the lognormal — presents problems, however. In a lognormal distribution the mode (the position of the peak on the density function plot or the maximum gradient on the distribution plot) and the median are coincident, but because of the extended right tail the arithmetic mean is displaced upward and is no longer at a position which would be intuitively identified as the 'centre' (Fig. 3.5). To take a more extreme example, in the exponential distribution, the mode is at a value of zero, the median is at some positive value, and the mean is at a higher positive value. A fourth case may be examined, of a bimodal distribution such as is frequently found in geochemical studies of mineralised areas. There is now no longer a single mode to use as a central estimate, and the mean and median show no obvious relationships with each other or with the form of the distribution. What, then, is the best general purpose estimator of the centre of a distribution? To try to answer this question it is necessary to examine the properties of available estimators in some detail. The mode may be eliminated immediately, as it is neither a stable nor necessarily unique value.

The properties of the mean are well known, from its central role in parametric statistics. The mean is the simple average value

$$\sum_{i=1}^{n} x_i/n$$

It is the value, for *any* continuous distribution, about which the sum of squares of deviations is minimised. Since the sum of squares

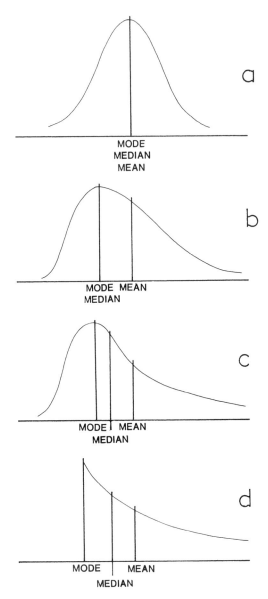

FIG. 3.5. Locations of the mean, median, and mode do not in general coincide. (a) In a unimodal symmetric distribution, they are coincident. (b) In a lognormal distribution, the mean is displaced from the location at which both mode and median occur. (c) In a more highly skewed distribution such as the gamma, the three measures all take different values. (d) In a negative exponential distribution, the mode is at the minimum, and the median and mean are both at higher values.

of deviations from the mean is n times the variance, the mean is therefore the minimum variance estimator of the centre of a distribution. However, as with all least-squares estimators, its value tends to be very sensitive to outlying observations because the resulting deviations *squared* have a disproportionate effect. In a lognormal or exponential distribution there is a long tail of high 'outlying' values without a matching low tail. The mean is thus displaced from what would otherwise be considered as the centre. Furthermore, the more extreme are the outliers, the fewer are required to perturb the mean value, which is thus unstable in the presence of long tails.

3.3.4.1 The Median

The population median may be defined in terms of the distribution as any value which bisects the distribution curve, such that exactly half the population lies below it, and half above. In a continuous distribution, the median will have a single value; in a discrete distribution (or a continuous distribution with gaps) it may be any value within a range (Fig. 3.4).

The sample median is an unbiased estimator of the population median, and is either the middle value (for odd n) or any value between the two middle values (for even n) — usually taken to lie midway between these values.

Whereas the mean is the estimate of the centre which minimises the sum of squared deviations, the median minimises the sum of absolute deviations

$$\sum_{i=1}^{n} x_i - M$$

Because the deviations are not squared, the median is affected much less than the mean by outliers, and for long tailed distributions is therefore more robust with respect to sampling of extreme values.

In any symmetric distribution the median coincides in location with the mean. In a normal distribution, the variance of the median is greater than that of the mean; its standard error is $1 \cdot 253\sigma/\sqrt{n}$ compared with standard error of the mean of $\sigma/\sqrt{n}$ (where σ^2 is the population variance). It is thus less good as a measure of central location than the mean, *for a normal distribution*. For long tailed symmetric distributions such as the double exponential this relationship is reversed. Indeed, in the Cauchy distribution, which has very high tails, there is no stable mean at all, as the value of $\bar{X}$ is controlled entirely by large values in the tails. In short tailed distributions on the other hand, both mean and median perform poorly at estimating central location, and the optimum sample statistic for the extreme case — the uniform distribution — is the mid-range (i.e. the mean of the lowest and highest observed values).

The median has some intuitive advantages over the mean: it divides the population into two equal parts; in skewed distributions it lies closer to the mode than does the mean; it minimises the sum of absolute deviations — easier to comprehend but mathematically less tractable than squared deviations. Its disadvantages, which have effectively relegated the median to a minor role in classical statistics, are in this mathematical inconvenience — because the sample median is not a simple analytical function of the data, the powerful methods of calculus cannot be used. Add to this the fact that it is less efficient than the mean for estimating the centre of a normal distribution, and it becomes quite clear why the mean has been preferred.

3.3.4.2 Trimmed Means
The mean and the median are in fact the two end-members of a class of estimators termed 'trimmed means'. The mean is the sum of all observation values divided by the number of observations. Any outliers will be included in the summation and will tend to move the mean away from its central position in the direction of the outliers. Moreover, in any distribution which has high kurtosis — one or both tails higher than in a normal distribution — the mean will tend to be unstable, its position controlled unduly by high and low values in the tails.

In such cases it would be desirable to improve the estimator by eliminating extreme values for the computation: if a few of the highest values are removed, together with the same number of the lowest, and the mean of the remaining values computed, the resulting estimator is the 'α-trimmed mean' where α is the proportion of values discarded at each end

$$m(\alpha) = 1/h \sum_{i = g + 1}^{n - g} x_{(i)}$$

where $g = n\alpha$ when α is selected such that g is an integer, or $g = U[n\alpha]$ in which $U[Q]$ is the greatest integer below Q; h is defined by $h = n - 2g$. If all values but the central one or two are trimmed, what is left is the median, which is thus a $\frac{1}{2}$-trimmed mean

$$M = m\left(\tfrac{1}{2}\right) = \tfrac{1}{2}(x_{(g + 1)} + x_{(n - g)})$$

where $g = U[\tfrac{1}{2}n]$.

Trimmed means with α between 0 and $\frac{1}{2}$ are sometimes used as robust estimators: for example, Tukey (1970) advocates the use of the 'mid-mean', the mean of the central half of the sample, which is a $\frac{1}{4}$-trimmed mean, $m(\frac{1}{4})$.

The properties of trimmed means are generally intermediate between those of the mean and the median, but without any of the 'nice' properties of either; however, they provide the basis for defining adaptive estimators, to be discussed below.

3.3.4.3 Adaptive Estimators

Adaptive robust statistics (Hogg, 1974) have been developed only recently, but they constitute a valuable alternative to the conventional approach, untrammelled by assumptions about the forms of distributions, though tailored to the observed distribution. The essence of adaptive statistics is that the estimators may be modified to match the form of the distribution as inferred from the sample observations. In the case of symmetric trimmed means, $m(\alpha)$, the amount of trimming, α, which is desirable may be altered and will depend on the kurtosis of the distribution. The normal parametric estimate of the sample kurtosis may by used to help define α, but Hogg (1972), Randles and Hogg (1973), Randles et al. (1973) have identified an empirically better indicator of the length of the tails

$$Q = [\bar{U}(p) - \bar{L}(p)]/[\bar{U}(0.5) - \bar{L}(0.5)]$$

with $p = 0.05$, where $\bar{U}(\beta)$ is the average of the largest $n\beta$ order statistics and $\bar{L}(\beta)$ is the average of the smallest $n\beta$ order statistics.

Q is a ratio of two linear functions of the order statistics and is therefore more robust than the parametric kurtosis estimate. Hogg (1974) suggests an even better indicator Q_1 for distributions with tails longer than those of the normal distribution, identical to Q but with $p = 0.2$. For samples drawn from symmetric distributions with tail lengths between those of the normal distribution ($Q_1 = 1.75$) and the double exponential ($Q_1 = 1.93$) Hogg uses the value of Q_1 to control the value of α in the trimmed mean $m(\alpha)$. For example, he suggests using

$$m(1/8) \text{ for } Q_1 < 1.81$$
$$m(1/4) \text{ for } 1.81 \leq Q_1 \leq 1.87$$
$$m(3/8) \text{ for } 1.87 < Q_1$$

For distributions close to normal this will give an estimator which is only a lightly pruned mean; for distributions with very long tails the estimator will be $m(3/8)$ which is nearly the median, having 3/4 of observations trimmed. It is, of course, possible to make the selection of α a continuous function of Q_1 if it is considered worth the effort for the slightly increased efficiencies which will be made possible.

For asymmetric distributions, Hogg suggests that an asymmetric trimmed mean $m(\alpha_1, \alpha_2)$ could be used as an adaptive estimator having α_1 and α_2 determined with reference to Q_1 (which indicates the length of tails) and Q_2, defined below, which is a measure of skewness

$$Q_2 = \left[\bar{U}(0.05) - m(0.25) \right] / \left[m(0.25) - \bar{L}(0.05) \right]$$

In asymmetric distributions, the *population* median and mean have different values, and the sample median and mean respectively are unbiased estimates of these parameters. However, the sample mean in a highly skewed distribution will be a very unstable statistic. In geostatistical studies, this has been recognised by de Wijs (1951, 1953) who proposed a formula for estimating the mean of a lognormal distribution from its median.

The adaptive asymmetric trimmed mean of Hogg will lie at some value between the mean and the median, depending on the kurtosis of the distribution, and will be controlled by the relationship which is defined between Q_1, Q_2, α_1 and α_2. However, if the asymmetric trimmed mean is an estimator of any parameter, that parameter is identified solely by this functional relationship. The concept of adaptive statistics is valuable but their usefulness and meaning will only be established if the adaptation of α_1 and α_2 is controlled by empirically *and theoretically* justified functions. For a generally acceptable 'centre' of any asymmetric distribution, then, it appears that only the mean and the median are available. Both are in use, in different circumstances; perhaps it may be relevant to look at the circumstances in which each is the appropriate parameter to use. Much of statistics has its roots in gambling, and this may be a suitable field in which to find an analogy. Consider a game of chance in which the win at each throw is a real value drawn from a positive skewed (e.g. lognormal) distribution. What is a 'fair' stake or entry fee for one throw? Of course, it depends on the interpretation of fairness. One definition, which is the interpretation made in classical statistics, is that the net winnings after a large number of throws be zero. Let the stake be set at v. Then after a large number of throws n, the total stake paid is nv. At the ith throw, the winnings will be w_i. The total winnings will then be

$$\sum_{i=1}^{n} w_i$$

and if the stake is fair the net winnings will be

$$\sum_{i=1}^{n} w_i - nv = 0$$

Rearranging this equation, it is found that the value of v must be the arithmetic mean of the distribution. There is, however, an alternative definition of fairness. If the *amount* of a win or loss is less important than the fact of winning or losing, then v should be set at a value which allows an equal chance of winning or losing. This value is the population median. It is the parameter to be used when the difference between $+ 100$ and $+ 200$ is less important than the difference between $+1$ and -1, as is indeed frequently the case in economic decision making.

3.3.5 Estimates of Dispersion
The conventional measures of dispersion are variance and its square root, the standard deviation. The variance is defined simply as the mean squared deviation about the mean, and is the value which is minimised by using the mean as the central parameter.

When using the median, however, it is the sum of absolute deviations which is minimised. The appropriate measure of dispersion here is thus related to the absolute deviations: both the mean and the median absolute deviation about the median are commonly used.

A hybrid measure of dispersion which has simplicity of computation but little else to recommend it is the mean absolute deviation about the mean: it has little relevance in parametric or nonparametric statistical theory and is not now in general use. The standard deviation and the mean or median deviation about the median are measures of the dispersion or spread of the distribution, and all are influenced to a greater or lesser extent by extreme values: the standard deviation and mean deviation about the median are both affected to about the same degree; the median deviation about the median is relatively unaffected and is thus the most robust of the three.

Many other estimators have been used, and some are optimal in special cases (for example, the range is the best estimator in the case of uniform distribution) but these are the only estimators to have gained wide acceptance.

3.4 TESTS

Tests are generally intended for use as confirmatory rather than exploratory statistics, though the boundary between the two objectives is fuzzy and tests may, for example, be used to provide information about the quality or reliability of estimators.

In the natural sciences, and particularly in geology, the role of exploratory statistics has been dominant, and relatively little attention has been paid to tests, with a few notable exceptions. The most important of these, in the context of this book, are the tests of distributions — much of the work of the statistician Kolmogorov, in fact, was inspired by distribution problems in geology. A little time should therefore be spent in examining the purpose of statistical testing.

Given a set of data about a class of events, a statistical test is a mathematical procedure which objectively allocates a number (usually interpreted as a probability) to a supplied hypothesis about the class of events, in the light of certain extra assumptions about the statistical properties of the parent population. The test is thus a convenient way to assess the reliability of extrapolating information from the known observed data to the entire class or population they are drawn from. As an example, a set of assays may appear to be drawn from a lognormal distribution; one or other tests may be applied to the data to determine the probability that this is indeed the case. A convenient terminology and notation used in statistical testing identifies the hypothesis whose probability of truth is being estimated as the 'null hypothesis' or H_0. If an alternative hypothesis is specified it is identified as H_1. Sometimes no specific alternative hypothesis is defined, in which case H_1 is the hypothesis that H_0 is false. In some tests it is possible to identify a number of different alternative hypotheses, H_1, H_2 etc.

There are a very large number of statistical tests in use, both parametric and nonparametric; it would be beyond the scope of this book even to summarise them. The reader is referred to the excellent texts by Siegel (1956), Gibbons (1971) and Snedecor and Cochran (1967). In this book only a few selected nonparametric tests will be examined which will be of use in geostatistical applications. The two classes of test which will be of particular relevance are tests for randomness of a data set, and tests of distribution.

3.4.1 Tests of Randomness
The tests discussed here are concerned with the null hypothesis that a sequential set of data are randomly distributed, against some alternative hypothesis such as that they follow a simple trend. There are two main groups of such tests: the first uses runs

of like values (or of increasing and decreasing values) and such tests are referred to generically as Runs Tests; the second uses rank-order statistics and adopts general techniques such as non-parametric correlation measures to test for randomness.

3.4.1.1 The Runs Tests

There are a number of different runs tests, all using similar ideas: a sequential data set will generally contain multiple occurrences of similar adjacent items (or may be transformed into one which does). For example, in a geochemical sampling programme the data of Fig. 3.6 were obtained. These are tabulated in Table 3.1. If the values are now transformed into a sequence of zeroes and ones — representing values below and above the median — they are in a form suitable for a runs test.

A run is defined as a set of similar adjacent observations. The maximum possible number of runs in a sequence of n items is thus n; the minimum number is two (given that there are two and no more than two possible data values). Either extreme case would be indicative of non-randomness in the data; between these is the average number of runs that would be expected if the data were random. It is possible to test for randomness against two different alternatives: either against the hypothesis of non-randomness, in which case either an exceptionally high or an

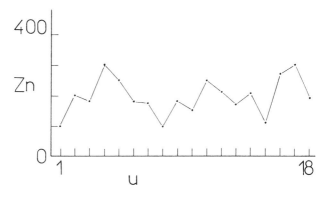

FIG. 3.6.   A sequence of geochemical data to be tested for randomness or trend. Data from Table 3.1.

TABLE 3.1

Geochemical data to be tested for randomness or trend. Indicator value is 1 for zinc above median, 0 for zinc below median.

Location	Zinc (microgrammes per gramme)	Indicator
1	100	0
2	200	1
3	180	0
4	300	1
5	250	1
6	180	0
7	175	0
8	100	0
9	180	0
10	150	0
11	250	1
12	210	1
13	170	0
14	205	1
15	110	0
16	270	1
17	300	1
18	190	1
Median value	185	

Maximum possible number of runs = 18
Minimum possible number of runs = 2
Observed number of runs   = 10

exceptionally low number of runs would cause H_0 to be rejected, or against the hypothesis of simple trend, in which case only an exceptionally low number of runs would cause rejection of H_0. For the two-tailed test, the null hypothesis of randomness would be rejected when

$$\left| \frac{R - 2n\lambda(1 - \lambda)}{2n^{1/2}\lambda(1 - \lambda)} \right| \geq z_{\alpha/2}$$

for a probability of rejection of α, where z_γ is the $(1 - \gamma)$ quantile point of the standard normal distribution, n is the total sample size, λ is the proportion of n which are observations of type 1 (the remaining $1 - \lambda$ being of type 2) and R is the total number of runs observed.

For many purposes, the differences between successive values are of interest: if all successive differences are computed from the data sequence, then runs of positive differences are in fact runs of increasing values, and runs of negative differences reflect decreasing values. A runs test using the successive differences (Edgington's test: Lewis, (1977, p.189)) is in fact equivalent to an ordinary runs test for $n - 1$ observations.

3.4.1.2 Mann's Test
Another test for randomness against the alternative hypothesis of trend uses the Kendall rank correlation coefficient, τ.

Mann (1945) studied the problem of such a test against the specific alternative of a downward trend (to test against an upward trend the data set need only be inverted or reversed). The principle is as follows: if there is a simple trend in a sequence of observed data, then the position of an observation in the sequence will show statistical correlation with the observed value at that position. In a data set whose randomness or trend is to be tested, it will therefore be sufficient to show the presence or absence of a significant correlation; a correlation coefficient may be used as the test statistic. Mann proposed that a suitable distribution-free correlation measure would be Kendall's τ, which is described below.

Correlation Measures
The traditional parametric statistic for estimating the correlation between two random variables X and Y is Pearson's product-moment correlation coefficient. This is computed directly from observed values and is thus a full parametric estimate. In recent years it has been shown by a number of workers to be highly sensitive to non-normality of the data distribution: in particular it is severely affected by outliers which have extreme values of both variables. There are, fortunately, alternative correlation measures available. The simplest of these is the Pearson correlation coeffi-

cient applied to ranks rather than to the data values. This is therefore obviously independent of the distribution; it is known as Spearman's rank correlation coefficient. Because (unlike the Pearson coefficient) the numbers to be used are known — sequential integers from 1 to n — it is possible to produce a very simple form of the correlation function

$$R = 1 - \frac{6\sum_{i=1}^{n} D_i^2}{n(n^2 - 1)}$$

where D_i is the difference between the rank of x_i and the rank of y_i. An alternative statistic, related to Spearman's coefficient, uses standard normal variates in place of the integer ranks; this coefficient was studied by Fieller et al. (1957). An unrelated coefficient, Kendall's tau, is sometimes preferred to these simple and intuitively attractive correlation measures, because its statistical properties are better known and more tractable. It uses the same amount of information as the Spearman coefficient, hence their efficiencies are the same: when used on data for which Pearson's coefficient may validly be used (bivariate normal) both have a power efficiency of about 91 percent (Siegel, 1956). Given two vectors of corresponding observations — say copper and zinc determinations in a batch of ore samples — the Kendall coefficient estimates the degree to which the two variables are associated or are independent of one another. The first step, as with many distribution-free techniques, is to transform the values in each vector to ranks.

Now, perfect correlation exists when, for every i and j, $x_i > x_j$ whenever $y_i > y_j$; the greater-than relationship is preserved under order preserving transformations, thus the values x and y may be replaced by ranks; perfect correlation may be defined as $r(x_i) > r(x_j)$ whenever $r(y_i) > r(y_j)$. To estimate the extent to which this condition is met in the observed data, it is necessary to examine this relationship for every pair of i and j values. This is numerically tedious, but is an easy operation. The $r(x)$ vector is ordered in increasing value of r, and the $r(y)$ vector rearranged to preserve the correspondence between matched x, y pairs. For every value of i from 1 to $n - 1$, the value of $r(y_{(xi)})$ (i.e. the rank of the y value that corresponds to x_j) is compared with every succeeding value of $r(y)$ in the rearranged $r(y)$ vector. If a value $r(y_{(xj)}) > r(y_{(xi)})$, $+1$ is added to a cumulative score; if on the other hand $r(y_{(xj)}) < r(y_{(xi)})$, -1 is added. For $r(y_{(xj)}) = r(y_{(xi)})$, the score is left unchanged. The principle is illustrated in Table 3.2. To obtain the tau coefficient, it is necessary to know the maximum possible score which could have been obtained. It is obvious that the maximum possible score is achieved when *every* $r(y_{(xj)}) > r(y_{(xi)})$ for every $j > i$. This maximum is thus $n(n - 1)/2$.

$$\tau = \frac{\text{actual score}}{\text{maximum possible score}}$$

hence, using S for the observed, actual, score

$$\tau = \frac{S}{\frac{1}{2}n(n - 1)}$$

It should be noted that, like the Pearson coefficient, tau may take any value between $+1$ for perfect correlation or concordance and -1 for perfect discordance or inverse correlation.

It is found in practice that although both Spearman's and Kendall's coefficients are based on the use of ranks, and use therefore the same amount of information from the data set, the numerical values generated by the two methods are different. This is because the measures of correlation are essentially different (though neither is more or less valid than the other).

To return to Mann's test for randomness against trend; a sequential data set may be considered as a bivariate sample, with one variable being the position in the sequence, the other, the observed value at that position. If the data show a simple trend, there is a significant correlation between these two variables. Mann (1945) developed the then recently introduced tau coefficient (Kendall, 1938) into a sensitive test for the null hypothesis of a random distribution against a simple downward trend; to test against an upward trend he proposed that the same test be applied using $-x_i$ values in place of x_i.

TABLE 3.2

Computation of Kendall's tau coefficient between zinc value and location for the data of Table 3.1.

Location	1	2	3	4	5	6	7	8	9	10	11	12	13	14	15	16	17	18
Rank	1	11	7	17	14	7	6	1	7	4	14	13	5	12	3	16	17	10

Indicator values: for location

	2	3	4	5	6	7	8	9	10	11	12	13	14	15	16	17	18
1	1	1	1	1	1	1	0	1	1	1	1	1	1	1	1	1	1
2		-1	1	1	-1	-1	-1	-1	-1	1	1	-1	1	-1	1	1	-1
3			1	1	0	-1	-1	0	-1	1	1	-1	1	-1	1	1	1
4				-1	-1	-1	-1	-1	-1	-1	-1	-1	-1	-1	-1	0	-1
5					-1	-1	-1	-1	-1	0	-1	-1	-1	-1	1	1	-1
6						-1	-1	0	-1	1	1	-1	1	-1	1	1	1
7							-1	0	-1	1	1	-1	1	-1	1	1	1
8								1	1	1	1	1	1	1	1	1	1
9									-1	1	1	-1	1	-1	1	1	1
10										1	1	1	1	-1	1	1	1
11											-1	-1	-1	-1	1	1	-1
12												-1	-1	-1	1	1	-1
13													1	-1	1	1	1
14														-1	1	1	-1
15															1	1	1
16																1	-1
17																	-1

Maximum possible score = 153
Actual score = 18
Kendall's tau = 0·12

3.4.2 Tests of Distribution

The classic test of goodness-of-fit between an empirically determined distribution and a theoretical distribution (or another empirically determined distribution) is the chi-square test. In most situations, however, this test is inferior in performance and flexibility to the family of tests devised by Kolmogorov and Smirnov, and tests related to them, such as that proposed by Sherman (1950).

The common aim of all such tests is to investigate the degree to which one distribution fits a particular 'model' distribution which is usually a theoretically derived distribution such as the normal or exponential.

The chi-square test, although an accepted part of 'classical' statistics, is in fact a nonparametric method; by its very nature it is distribution-free. Data must be supplied in grouped form — exactly as produced when plotting a histogram. The sample space is divided into a number of classes into which the data are categorised.

Now, if a random sample of the same size were drawn from a population having the ideal 'model' distribution, its 'observations' could also be classified into the same classes. The chi-square test compares the numbers of real and simulated data recorded in each class: the higher the discrepancy, the less the probability that the two distributions are in fact the same. The chi-square test statistic U uses the squared discrepancy to ensure a positive value. U is defined as

$$U = \sum_{i=1}^{k} \frac{(n_0(i) - n_t(i))^2}{n_t(i)}$$

where $n_0(i)$ is the observed frequency for class i, $n_t(i)$ is the theoretical or expected frequency for class i. The distribution of U is approximately that of chi-squared (a distribution closely related to the normal) — hence the name of the test.

The main problems of the chi-square test lie in its use of grouped data — it is not suitable at all for use with small samples and becomes unreliable when any class has an expected frequency below five.

3.4.2.1 The Kolmogorov and Smirnov Tests

The approach of these tests is quite different from that of the chi-square. Instead of using grouped data — and thus loosing a large amount of information present in the individual observations — these tests use every observation. The general approach is to plot (or at least to compute) the empirical distribution curve (Fig. 3.7) and also to plot the theoretical or observed distribution with which it is desired to make a comparison.

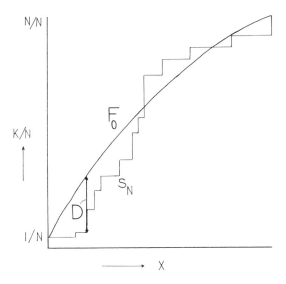

FIG. 3.7.   Theoretical (F_0) and observed (s_n) cumulative distribution curves for a set of data. Maximum vertical deviation between them, D, is the Kolmogorov–Smirnov statistic.

If the two distributions can reasonably be supposed to be identical, then for every value x, $s_n(x) = k/n$ (where k is the number of observations less than or equal to x) should be fairly close to $F_0(x)$, the 'ideal' value of the distribution function for value x. The Kolmogorov–Smirnov and related tests concentrate on the largest deviation, that is the largest value of the difference between $s_n(x)$ and $F_0(x)$:

$$D = \text{maximum}|F_0(x) - s_n(x)|$$

The probability distribution of D does not depend on either the observed distribution or the theoretical distribution, and the technique is thus truly distribution-free.

It is also possible to define one-sided Kolmogorov–Smirnov statistics D^+ and D^- which record respectively maximum positive and negative deviations from the theoretical distributions. Again the distributions of these statistics are independent of the theoretical or observed data distributions.

There are two main ways in which such statistics may be used:

(i) to compare two empirical distributions (Smirnov-type statistics); (ii) to compare an observed with a hypothesised distribution (Kolmogorov-type statistics). The various ways these are used are explained very well by Lewis (1977, pp. 212–26), Siegel (1956, p.4752) and Gibbons (1971, pp.75–87); a further extensive treatment seems unwarranted because of the simplicity of the concepts involved. Lindgren (1976) discusses not only the Kolmogorov–Smirnov test statistic, but also some related methods (such as that proposed by Cramér and von Mises (Darling, 1957): an integral function of squared deviations) and a much simpler function suggested by Sherman (1950). Sherman's statistic is

$$n = \frac{1}{2} \sum_{i=1}^{n+1} \left| F_0(x_{(i)}) - F_0(x_{(i-1)}) - \frac{1}{n+1} \right|$$

where $x_{(0)} = -\infty$ and $x_{(n+1)} = +\infty$. The reasoning behind this statistic is that the expected area of the density curve between a pair of successive ordered observations is $1/(n+1)$ but the area that would be produced if F_0 were the correct distribution would be $F_0(x_{(i)}) - F_0(x_{(i-1)})$. It therefore measures a total area discrepancy between the two distributions. Sherman's statistic is distribution-free since it operates on the order statistics.

3.5 APPLICATIONS

Many nonparametric and distribution-free statistical methods have been omitted from this chapter — those selected were those of most relevance to the techniques to be developed in Chapter 4. However, many nonparametric methods are of direct application to a wide range of problems in geology and the other natural sciences. Some have already been used, but few such methods have been readily accepted as the standard, preferred method, even though it has become increasingly obvious that the traditional methods, straitjacketed by normal distribution theory, requirements of homoscedasticity, and many similar assumptions, are not altogether appropriate to sciences in which simple linear interactions are exceedingly rare occurrences. Lewis (1977) describes the advances that have been made by nonparametric techniques in geography; there is scope for extension of their application into many related sciences.

CHAPTER 4

Nonparametric Geostatistical Estimation

$\pi \acute{a}\nu\tau a \ \dot{\rho}\varepsilon\hat{\imath}, \ o\grave{\upsilon}\delta\varepsilon\nu \ \mu\acute{\varepsilon}\nu\varepsilon\iota$
'All is flux, nothing is stationary'
HERACLITUS

4.1 PRELIMINARIES

The reader will probably have seen no necessary connection between the contents of Chapters 2 and 3. Nevertheless, most of the nonparametric principles and methods introduced in Chapter 3 may be used to ease the constraints imposed on geostatistical techniques by assumptions of distribution and spatial homogeneity.

Two assumptions, in particular, are irksome, and have led to the development of more complex geostatistical estimation methods, in attempts to circumvent them. These are:

(i) that the data are drawn from a normally distributed population (or at least one which can be transformed simply to a normal distribution);

(ii) that the data are drawn from a spatially homogeneous population in which the distribution is independent of location in space and whose autocovariance relationships may be expressed in the form of a 'variogram' or 'semi-variogram' which is invariant under translation space.

A third assumption, pointed out by Journel and Huijbregts

(1978), but usually ignored in geostatistical studies, is that the variable studied must be capable of additive relationships. This point is treated in greater detail in Section 4.2.4.

In contrast, one of the great advantages of nonparametric statistical techniques is the laxity of conditions required of the data. There is usually little constraint on the distribution (for example, only continuity is assumed, rather than normality), and care has been taken in developing the methods introduced in this chapter, to ensure that continuity is the only assumption made about spatial relationships, rather than *any* form of stationarity; neither autocovariance (and variogram) nor distribution are assumed to remain constant with translation in space.

It will be noted, however, that parametric geostatistics is concerned principally with exploratory statistics and estimation rather than confirmatory statistics and tests. The nonparametric statistical techniques used most widely are concerned with testing of hypotheses. The chapter will show how the principles and ideas used in such tests may be adapted for the purpose of estimation of spatially distributed variables.

4.2 ASSUMPTIONS

The purpose of nonparametric geostatistics is to provide an estimation tool with as great a range of validity as possible, while assuring a solution which is optimal in the sense of minimising absolute deviations. The assumptions required must therefore be as few and as unrestrictive as possible.

4.2.1 The Parent Population

Consider first the population or universe from which a sample is drawn. This population may be defined as the sum of a smoothly varying function and a random component. This definition is not that of a 'regionalised variable' of Matheron's geostatistics in which the term 'smoothly varying' implies the presence of trend, considered a nuisance, and poses problems in estimating the semivariogram (indeed the general problem of elimination of trend effects in parametric geostatistics has not yet been solved). However, without such smooth variation — on an undefined scale, and probably on a range of different scales — there would be no scientific phenomena to study, and the best estimate of the value of the variable at any point would be the global mean.

The random component is assumed to be a point function and to be related mainly to the observational technique: it is the observational error, and in parametric geostatistical language is the pure 'nugget effect'. Any other small-scale variation is, by definition, a variation of real values and is part of the smoothly varying function, on any scale above that at which quantum effects predominate. Thus an actual gold nugget will contribute a steep sided peak to this function but will not contribute to the pure nugget effect. There may well be views opposed to this simple model of observed spatial data. For example, Matheron's regionalised variable is a powerful concept and yet it is a purely statistical one, with no simple interpretation in terms of geological surfaces or other data.

One geologically useful class of functions at present is excluded from the population definition above; that is, those functions which are continuous *almost* everywhere but display isolated cusps or discontinuities and may also be locally multi-valued. Such are the surfaces which are of interest to the differential topologist, and are described by Catastrophe Theory (Thom, 1975); geological examples are many, and are generated by faults and by a variety of physically discontinuous processes (Henley, 1976; Cubitt and Shaw, 1976). Work is in progress to extend nonparametric geostatistics to include such surfaces — some of the relevant ideas will be introduced in Chapter 6.

4.2.2 Homogeneity

Although it is required by parametric regionalised variable theory (at least in the simpler generally accepted forms), the assumption of statistical stationarity is neither required nor adhered to by the processes acting in geology to generate the observed phenomena. Values generated by these processes are determined purely by the mechanisms of the processes and by the complex interactions between different processes acting simultaneously. Natural processes are not constrained to produce resultant mineral grade porosity, grain size, or other properties with the desired stationary numerical behaviour.

For nonparametric geostatistics, therefore, such stationarity assumptions are not made, and the only requirement of spatial homogeneity is that the observed property be any wholly continuous function of location in space (it will in general *not* be a function which is described by any mathematical expression, but may be considered a smooth function in the topological sense). Such a homogeneous region is bounded only by major observable boundaries (i.e. faults) or the limits of the region under study. It is not necessary for this smooth function to be observable directly — indeed, the observation–error nugget effect prevents any possibility of this. It therefore shares with the regionalised variable of parametric statistics the property that its 'real' structure is unknowable.

4.2.3 Distribution
Because the parent population (excluding for now the nugget effect) is largely deterministic, even if unknown, it is not strictly valid to speak in terms of 'probability distributions'. However, the data values may still be considered as being drawn at random from a parent distribution which is determined by the nugget effect error distribution and by the functional form of the whole population. In general, given a continuous parent population, the distribution will also be continuous.

However, the distribution usually cannot be normal or lognormal: both distributions include values which tend to infinity. The data distribution will, in fact, always be bounded above and below by the physical limits on values attainable by the observed property. For example, the amount of copper in an ore sample cannot be less than zero, and cannot exceed 100 percent. Although there are statistical 'model' distributions available which are bounded in this way, there is no justifiable reason to select any particular one of these and use it as a model of the data. It is much more reasonable, in view of the population model defined above, to leave the distribution undefined beyond the statement that it is continuous.

4.2.4 Additivity

It is *not* an assumption of nonparametric geostatistics that a variable should possess the property of additivity, as is required by parametric geostatistics (but is frequently ignored in studies which use kriging methods on unsuitable variables). There are many variables whose spatial variation one might wish to study, that are not validly handled by methods based on parametric regionalised variable theory, but for which no suitable valid estimation method has yet been available.

Such properties as temperature, porosity, permeability and resistivity are not additive and are not properly handled by kriging methods (Journel and Huijbregts, 1978, p. 199) yet in most cases parametric geostatistics do not provide an alternative method which is valid. In some cases when an alternative is available, great care must be taken in using it. For example, metal grades in an ore-body may be kriged. So may accumulations (measured in a layered gold deposit, for example, as inch-dwt or gramme-cm). Because these accumulations are recorded in such hybrid units, however, they are not directly additive and must be converted into a standard form before kriging. The inch-dwt unit multiplies the number of dwt (of gold) per ton (of ore) — a grade — by the thickness of the seam in inches, and cannot be kriged directly but must be multiplied by the density to yield the number of dwt per unit area of the deposit (a true accumulation value), which is an additive variable. Although such a transformation is valuable in any case as it is an aid to clear thinking, it is not one which is required for validity of nonparametric geostatistics.

4.2.5 Summary

The assumptions which are to be used in nonparametric geostatistics may be listed as:

(1) The parent population or universe has a probability distribution which is continuous but may vary in form from one

location to another: in particular, its centre and its dispersion are both expected to vary.

(2) All differences between observations at different locations may be attributed to a combination of only a smoothly varying function of location and a point random error (a 'nugget effect' in the language of parametric geostatistics), which may itself have any continuous distribution.

(3) There is no assumption of stationarity and no need for the concept of stationarity of any order. If a variogram is computed, its form is a reflection of the local form of the smoothly varying function which defines the surface, and of the point random errors.

(4) This smoothly varying function, which combines the properties of 'trend' and non-nugget residuals of parametric geostatistics, is emphatically not interpreted as a simple mathematical function, but as the *actual* geological (or other) surface whose values are sought.

(5) There are only two random terms involved in estimation of this surface; these are the point error term, and the uncertainty due to lack of full information on the form of the function, caused by finite spacing of the pattern of observations.

4.3 THE NONPARAMETRIC ESTIMATION MODEL

As described in Chapter 2, the linear estimation model in its simplest form is

$$T^* = w_1 g_1 + w_2 g_2 + \ldots + w_n g_n$$

If all w_i are equal (at a value of $1/n$) then T^* is the arithmetic mean of all observations g_i. If T^* is an estimator of a true value T then the estimation error is $\varepsilon = T - T^*$. T^* is an unbiased estimator of T, and $E(\varepsilon) = \bar{\varepsilon} = 0$. If all g_i are drawn from the same distribution, then this is also the minimum variance estimator — $E(\varepsilon^2)$ is minimised. In geostatistical problems in which the g_i are observations at different spatial locations, the individual w_i for $E(\varepsilon^2)$ to be minimised are unequal, but the model is essentially the same.

The linear model will in general provide an estimator with which once can minimise the sum of *squares* of deviation between estimated and true values. Such estimators are very commonly used, because of the relative simplicity of the mathematics. However, it is intuitively at least as good to search for an estimator which minimises the sum of *absolute values* of deviations. Just as the unweighted arithmetic mean is the simplest case of a linear estimator, so the sample median is the simplest case of this second type of estimator; we may generalise the median by introducing the concept of a *weighted median*

$$M^* = \text{med}\,(F_w\,(g))$$

where F_w is a weighted distribution function in which each g_i value is assigned a relative frequency w_i and

$$\sum_{i=1}^{n} w_i = 1$$

Figure 4.1 shows a simple example of a weighted median, compared with an unweighted median for the same set of observations. The weights would reflect variations in relevance and reliability among the observations.

4.3.1 The Moving Median

In the interpolation or gridding of spatial data, one of the simplest and most commonly used methods is that of moving averages. A 'true value' x_p at point P is estimated by taking an average of all surrounding points within a certain limiting radius (sometimes, for computational convenience, even with a surrounding square or rectangular area). This average $\bar{x}_p$ is a linear model estimator with weights $1/n$ on the n points lying within the radius, and 0 on all other observations. Since no attempt has been made to optimise these weights as pointed out by Clark (1979) the resulting estimate of x_p does not necessarily minimise the sum of squared deviations.

A direct analogue to this method would be to compute a simple

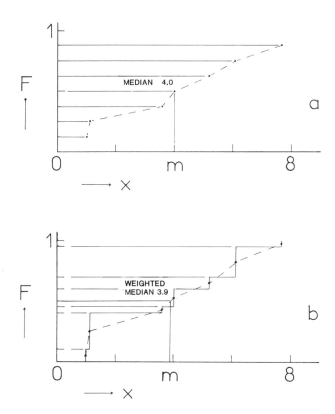

FIG. 4.1. Graphic demonstration of the method of estimating unweighted and weighted medians. Ordered data:

Value	1·0	1·1	3·6	4·0	5·2	6·1	7·7
Weights(a)	1	1	1	1	1	1	1
Weights(b)	0·1	0·3	0·05	0·15	0·1	0·25	0·05

median of the n observations which lie within the limiting radius. Again because the weighting system is crude, there is no indication of the quality of such an estimate, and one would expect it to be rather poor.

In either case, one of the reasons why the estimator is not very good is that the value at the centre of the search circle is being estimated from observations that lie close to the limiting distance as well as from observations very close to P, with equal weightings.

In fact, it would generally be expected that observations close to P would be better guides to the value at P than observations at points further away. In moving average interpolation, therefore, it is common for some sort of distance weighting to be used, in which points beyond the limiting radius are still allocated a weight of zero, but the weights given to points within it depend on the distance of the observation from the central point P. Weightings which are conventionally used include inverse powers of distance, negative exponential powers of distance (e.g. e^{-d}, e^{-d^2}) and a variety of other mathematical functions. All share the property that the weighting function is a monotone decreasing function of distance. The kriging methods take this concept a good deal further by taking into account the relationships between variance and distance, as described in Chapter 2, and generating weightings for each observation, for each point P to be estimated; these weightings reflect not only the distance between the observations and P but also the distances between the observation and other surrounding observations.

A *weighted moving median* is a direct analogue of a weighted moving average; weights may be computed in the same way, as functions of distance — indeed, kriging weights may also be used if desired, though they are not altogether appropriate as they were themselves generated on the assumptions of a linear model. Since the notion of a moving median is altogether new, some time will be spent in examining its properties.

4.3.1.1 The One-Dimensional Case
Consider a one-dimensional set of data — a sequence such as a time-series or a geophysical traverse. The analogue to the search circle is a window of a given length $2r$, which defines a neighbourhood of point P (at its centre) within which observations are to be given non-zero weightings. The population of true values will consist of a set of points forming a smooth curve of some sort, of unknown degree of complexity. Given a small enough window, however, this complexity may be reduced so that the population lying within it is one of three types: it may be a simple monotone curve, it may be a curve with a maximum, or it may be a curve with a minimum (Fig. 4.2).

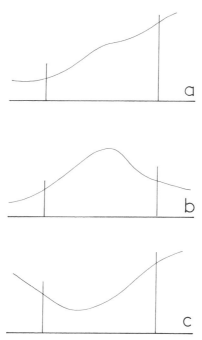

FIG. 4.2. The possible shapes of curve segments in one dimension. (a) Monotone, increasing or decreasing; (b) a maximum; (c) a minimum.

To examine the simplest case first: a simple monotone curve lying within the window. It will be readily apparent that in *any* such case, the true value at *P* is the population median (Fig. 4.3), since all values on one side of it are lower, all on the other are higher, and the distances within the window on the two sides are equal. It is important to note that the population mean will not generally coincide with the true value at *P*, whereas the median will. A set of random observations lying within the window may be used to estimate x_p; the sample median will always be an unbiased estimator, but the sample mean will almost always be biased. The same applies to distance weighted estimates: the weighted median is unbiased, the weighted mean is biased.

It is true, of course, that over the universe of all possible monotone curves within the window $P - r$ to $P + r$, both the mean and the median are unbiased estimators of x_p. However, in the universe (rather fuzzy) of 'reasonable' curves as conditioned by the

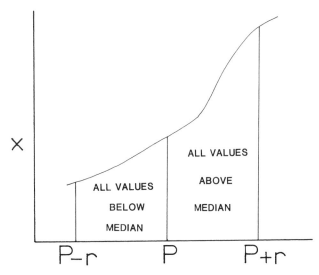

FIG. 4.3. The value of the point at the centre of a window on a monotone section of curve is the population median of the values within the window.

data, the median remains unbiased while the mean will usually become biased.

No such statements can be made when the neighbourhood of P contains a maximum or a minimum (Fig. 4.4). The median is the value above which lies one half of the values in the interval $P - r, P + r$; x_p will not generally coincide with the median when this interval includes a maximum or a minimum. However, the population mean will tend to be an even worse estimator of x_p. An unbiased estimator of x_p would be the qth quantile, where true values lower than x_p occur over a proportion q of the interval $P - r, P + r$.

This statement may be generalised to cover not only the simple monotone curves and the cases of single maxima or minima but also more complex 'real world' situations. x_p may be defined exactly (for a continuous curve) as the qth population quantile, of the population defined as that part of the curve lying within the interval $P - r, P + r$, where a proportion q of values lie below x_p, or equivalently over a total distance $2rq$ the curve consists of

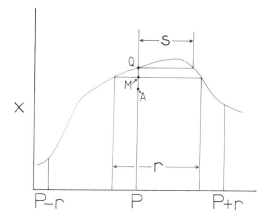

FIG. 4.4.   If the window (of length $2r$) contains a maximum, the value at the centre P is not the median M or the arithmetic mean A, but some quantile Q above the median. The proportion of the population lying below Q is $(2r - s)/2r$.

values smaller than x_p (and over distance $2r(1 - q)$ the true values are greater than x_p).

In any interval in which the curve is in fact monotone, q will take the value of $\frac{1}{2}$, and x_p will be the median.

Given a sample consisting of evenly spaced observations in $P - r$, $P + r$, the task of estimating x_p may alternatively be expressed as the problem of estimating q. It will be seen immediately that no simple linear combination of the observations x_i (as some form of weighted average) is suitable for this purpose. This includes kriging estimators. An unweighted median of x_i observations lying within $P - r$, $P + r$ is better, but is insensitive to variations in spacing of the observations (Fig. 4.5). A weighted median may be used, with weights w_i defined as some inverse function of distance such as $1/d_i$; this is less sensitive to the spacing of observations and in the case of a monotone curve in the neighbourhood of P it provides a good estimator of P: it is almost equivalent to a simple linear interpolation between the two observations closest to P (on opposite sides). Use of a median, weighted or unweighted, however, implies a q value of $\frac{1}{2}$, and so parts of curves which include maxima or minima will not be well fitted. A possible

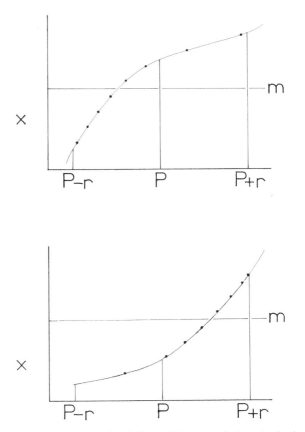

FIG. 4.5.   An unweighted median is insensitive to variations in the locations of
observations within the window.

approach to this problem will be examined in Section 4.3.2, in
which a fairly simple way of estimating q is suggested.

4.3.1.2 Two and Three Dimensions
The problem in the two-or three-dimensional case is the estimation
of a point value at the centre of a neighbourhood which consists
of a portion of surface or hypersurface rather than a simple smooth
curve. It is still assumed, however, that the surface or hypersurface
is a continuous function. The following discussion will concern the

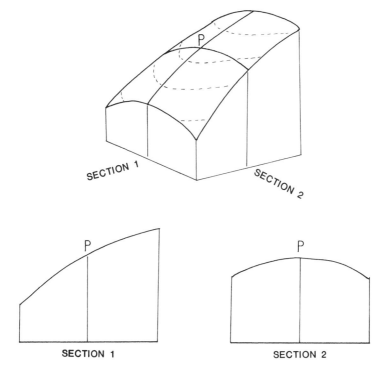

SECTION 1   SECTION 2

FIG. 4.6. A two-dimensional observation space cannot be considered as a simple extension to the one-dimensional case. Some sections through P (e.g. section 1) may be monotone, while others (e.g. section 2) are not.

two-dimensional case: the same arguments are equally valid for three dimensions.

Consider a point P surrounded by a circular neighbourhood of radius r. Within this circle, one half of the population of x values will lie above the population median, one half below, by definition. The concept of a monotone curve is meaningless, however, since in more than one dimension only partial ordering is possible; it can be seen from Fig. 4.6 that even a simple smooth surface without maxima or minima may include one-dimensional sections which pass through P and which *do* include maxima and minima. It is thus not possible simply to extend one-dimensional conclusions into two or three dimensions. In general, therefore, it cannot be assumed that x_p is the population median, even when the surface contains no maxima or minima within the neighbourhood of P. The median is still of some value, however. Given an area A of size πr^2 and an arbitrary point i within it of value x_i, the probability that x_i exceeds the population median M_A of A is the same as the probability that x_i is less than the median, since the areas of values above and below the median are equal:

$$p(x_i > M_A) = p(x_i < M_A) = \tfrac{1}{2}$$

The absolute value of the deviation of x_i from the median is

$$D_i = x_i - M_A$$

Now, since the median is the central estimator which minimises the sum of absolute deviations, it can be shown that M_A is the value about which absolute deviations are least. Thus, given a *sufficiently* small neighbourhood area A, the median value within this will, on average, lie closer to any arbitrary value within it — including the value at the central point P — than any other unbiased estimator (such as the mean). M_A is the minimum deviation estimator of the centre, as contrasted with the mean, which is the minimum variance estimator.

In the same way that the optimum sample statistic to be used to estimate x_p with minimum variance is in fact a weighted mean, to allow for the differing relevance of observations at different distances from P, so a weighted median must be used to give an optimum minimum deviation estimate. This accounts for the emphasis above on the words 'sufficiently small'.

4.3.1.3 The Geometric Problem

As can be seen in Fig. 4.7, x_p is not generally the median of population A, even when the surface is of simple form (Fig. 4.7(a)). However, when A includes one or more maxima or minima, the situation is no more complex (Fig. 4.7(b) and (c)). In each case, a proportion q of the area contains values below x_p, the remaining $(1 - q)$ proportion of A contains values above x_p. The two areas are separated by a line which is the contour line or *isoline* connecting all points which have value $x = x_p$.

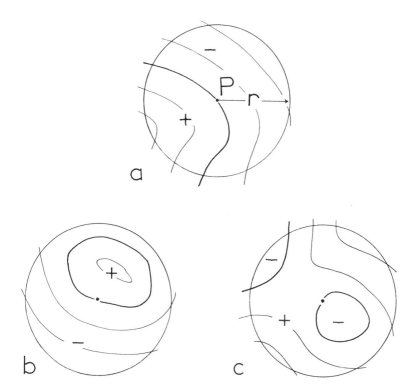

FIG. 4.7. The value at P is not in general the population median of a two-dimensional window. (a) The areas of values above and below the value at P are in general unequal. (b) The situation is however no more complex if a maximum or minimum is included within the window. (c) Areas of values above or below P may be disjointed.

It should be remembered that the values being used here are *population* values, and are in fact unknown. The form of the surface within A must be estimated from whatever observations are available.

4.3.2 The Varying Quantile Method
In order to obtain an estimate of the value x_p which is better than the median in one-, two-, or three-dimensional cases it is necessary to be able to estimate q, the order statistic (or weighted order statistic if the contributions of observations are weighted) which

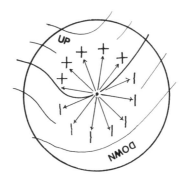

FIG. 4.8. P may be considered as the centre of an area within which the gradient
of the observed value in any given direction is either upward or downward away
from P. This property may be used to provide estimates of the areas above and
below x_p.

may be used with the set of observation values, weighted or unweighted, to determine the estimate $\hat{x}_p$.

Point P may be considered as the centre of an area within which,
in any direction, the gradient of x value is either upward or downward away from P (Figure 4.8). Now, if (within a small neighbourhood of P) the observed values x_i show a regular increase away
from P in *all* directions, the best estimate of q is zero; x_p is a minimum on the surface, and there is no area (within the neighbourhood) of value below x_p. Similarly, if the observations show a
regular decrease in value away from P in every direction, then the
best estimate of q is 1, and x_p is interpreted as a maximum. In
intermediate cases, there will be some evidence for an increase
in value away from P, and some evidence for a decrease; an estimate of q in such cases will lie at some value between 0 and 1.

To formalise this approach a little, there are some standard statistical procedures which are used for testing the hypothesis of
'trend' against an alternative hypothesis of absence of trend or statistical randomness (this alternative hypothesis could be interpreted as an estimate $q = \frac{1}{2}$). Some of these tests are described by
Lindgren (1976), Gibbons (1971) and Siegel (1956). They include
the runs tests and Mann's test, as introduced in Chapter 3. To
apply such tests to spatially distributed data, the observations

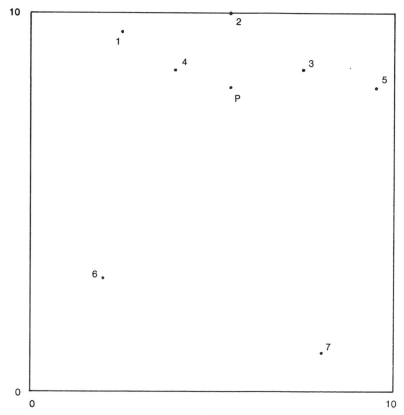

FIG. 4.9. Locations of sample points and point P whose value is to be estimated, for the data of Table 4.1.

must be ordered into a sequence of increasing distance from P within the neighbourhood. For the moment it will be assumed that all weightings are equal, and the question of definition of size of the neighbourhood (radius r) will be left unanswered.

The method will be illustrated by a geological example (Fig. 4.9) in which a number of copper analyses of samples collected irregularly are to be used to estimate the copper value at point P. The data are listed in Table 4.1.

Mann's test will be used to produce an estimate of q. The data are first ordered into increasing distance from P; each analysed value x_i is then compared with each more distant value x_j and an

TABLE 4.1

Synthetic geochemical data: copper values in seven point samples from a two-dimensional sampling space illustrated in Fig. 4.9.

Observation number	Copper (microgrammes per gramme)	Easting	Northing
1	210	2·5	9·5
2	300	5·5	10·0
3	450	7·5	8·5
4	500	4·0	8·5
5	430	9·5	8·0
6	260	2·0	3·0
7	290	8·0	1·0
P[a]	—	5·5	8·0

[a]Estimate required at location P.

indicator of $+1$ recorded if $x_j < x_i$, -1 if $x_j > x_i$ (Table 4.2). Given n observations within the neighbourhood, the maximum possible number of $+1$ (or -1) indicator values that can be recorded is $\frac{1}{2}n(n-1)$. The recorded indicators $+1$ and -1 are summed (the case of $x_i = x_j$ is recorded as a zero), and divided by $\frac{1}{2}n(n-1)$

TABLE 4.2

Indicator values for Mann's test, computed from the data of Fig. 4.9 and Table 4.1 (compare with the method of computation of Kendall's tau illustrated in Table 3.2).

Distance from P	Observation number	Indicator values for observations:					
		4	2	3	1	5	6	7
1·58	4	1	1	1	1	1	1
2·0	2		−1	1	−1	1	1
2·06	3			1	1	1	1
3·35	1				−1	−1	−1
4·0	5					1	1
6·10	6						−1
7·43	7						

to obtain Kendall's tau. This will thus take some value between
+ 1 and − 1. Now a value for τ of + 1 implies that all more distant
values are lower than all closer values, and so $q = 1$; a τ of -1
implies the reverse, that $q = 0$, and a τ of zero implies $q = \frac{1}{2}$, since
there is no demonstrated trend in values related to distance from
P. To derive an estimate of q from τ it is thus necessary to transform the range of τ (-1 to $+1$) into the range of q (0 to 1). τ and
$\hat{q}$ are most simply related by the equation

$$\hat{q} = \tfrac{1}{2} + \tfrac{1}{2}\tau$$

reflecting the default state of $\hat{q} = \frac{1}{2}$ in the absence of information
about the trend. A non-zero τ value is used to add bias to the
estimate (which would otherwise simply be the median) and allow
an estimate which is conditioned to the observed values. Use of the
Mann/Kendall method thus allows an adaptive statistical approach to be made, similar in spirit to that adopted by Hogg, as
described in Section 3.3.4.3. It should be emphasised that estimates obtained from such a procedure, while individually providing minimum deviation unbiased point values, are not
guaranteed to give a smooth estimated surface. This point is
expanded upon in Section 4.5.2.

4.3.2.1 Distance Weighting

The use of equal weightings for all observations within a radius r
from point P, and zero weighting for all other observations, is,
as has already been pointed out, a very crude weighting scheme,
in which points close to the boundary of the neighbourhood
may either be considered fully representative members of it, or
not considered members at all. Much more reasonable is a system
of weights in which every observation in the data set is considered to have some probability of 'belonging' to the neighbourhood of point P; the farther away from the point from P, the less
this probability. This concept produces a distance weighting
scheme of some sort, in which each observation is allotted a
weight which is a function of its distance from P. The observation
closest to P must obviously be included in any estimation, and is
thus given a probability, or a weighting, of one. The simplest form

of distance weighting which can then be used is of the form $w_i = d_{min}/d_i$, where d_{min} is the distance from P of the closest observation. This is almost equivalent to simple inverse distance weighting, with one important exception; for $d_{min} = 0$, the closest observation is actually at P, and the weights on all other points are set at zero: this effectively forces the method to fit data points exactly.

There is, however, a big drawback in using inverse distance weighting: if data points are scattered at approximately constant density over the whole sample space, then in two or three dimensions there will be a greater contribution to the estimate from distant observations than from closer observations. This point is shown by Fig. 4.10, for a two-dimensional case. There is one observation at distance 1, two at distance 2, each with a weighting of 1/2, and three at a distance of 3, with a weighting of 1/3. The total of weightings at each radius is the same. In general, the number of observations in any annulus between distances d and $d + \delta d$ is proportional to d in a two-dimensional sample space, and is proportional to d^2 in a three-dimensional space. Thus, for two-dimensional sampling, a weighting of d_{min}/d_i leads to an unstable estimate because distant observations have a larger cumulative effect than closer observations. An alternative weighting function which is very commonly used for moving averages is inverse distance squared (or inverse squared distance, or isd). This may be redefined analogously to the inverse distance function defined above, as

$$w_i = (d_{min}/d_i)^2$$

Of course, this weighting function itself leads to unstable estimates in a three-dimensional sample space. A stable estimate in three-dimensional space requires a weighting function which falls off with distance faster than the inverse square; for an acceptable weighting scheme

$$\frac{dw}{dd} < -2d^{-3}$$

Such a weighting function could be an inverse power, where the power is set by the dimensionality of the sample space. Given a

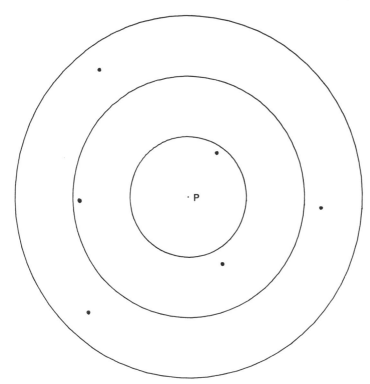

FIG. 4.10. In a two-dimensional sampling space, with evenly spaced observations, the number of observations increases in proportion to the first power of distance away from P.

sample space of dimensionality k, the weighting function is

$$w_i = (d_{min}/d_i)^k$$

Thus for 3 dimensions it would be an inverse cubed distance.

This is not necessarily an optimal weighting function for any particular set of observations, but it is one which will at least be certain to provide a stable estimator.

4.3.2.2 Range

In parametric geostatistics, a range beyond which observations are individually considered to have no influence on the estimator is determined by interpretation of the variogram: if the variogram appears to fit a transitive model, then the range is defined by the point of transition. However, two variogram models of wide application do not in fact show transitional behaviour: these are the exponential and the logarithmic (de Wijsian) models. When either of these is found in parametric geostatistical studies, a range is chosen as a matter of convenience rather than from some theoretical reasoning. In the same way, in nonparametric geostatistics, a range — the radius of the search area — may be chosen for convenience, at a distance beyond which the weightings on each observation can be considered to be insignificant.

4.3.2.3 Independence of Observations

When the spacings between observations are uniform, as in a grid sampling pattern, every observation carries the same quantity of information as every other. However, when some observations are clustered together in space, the information they contain is partly duplicated. In an extreme case, ten (say) replicate observations at the same point in space will each contain the same amount of information, but together they will not give any more information about the form of the surface than would one single observation.

Thus a weighting factor of 1/10 should be applied to each observation to indicate its degree of relative independence from all other points. Such a weighting is implicit in kriging techniques, where weightings are computed according to the covariances between observation points; however, kriging weights must be computed (for an irregular sampling pattern) separately for every observation used, for every point being interpolated.

The geometry of the pattern of points is the sole factor which affects the amount of information contained in the observations; the relevance of this information to any particular point being estimated is determined by the distance weighting function and is all that should need to be separately computed at each interpolation point. It therefore ought to be possible to generate, once only for a given set of observations, a set of 'independence' or 'cluster' weightings.

Now, a cluster of observations grouped more closely than the average sampling density represents a deviation from the theoretical uniform random point distribution which would generate equal weightings on all observations. The degree of deviation of a sample from a theoretical distribution may quite readily be estimated using a standard statistical technique such as the chi-square test or the Kolmogorov–Smirnov test. In the present case, the Kolmogorov–Smirnov test would seem to be the more appropriate since it considers observations individually; the chi-square test would require the data to be grouped in some way.

The problem is to determine, for each observation point, the degree of deviation (in the direction of clustering only, since it does not matter if observations are anti-clustered) from an ideal random distribution in space. Observation point density is related to distance, area, or volume depending on the dimensionality of the sample space.

The distances of observations $i = 1, 2, \ldots, N$ from observation P whose 'cluster weight' $w_c(P)$ is to be determined, are first converted to area or volume if appropriate, and sorted into increasing order of a_i (where a denotes the measure of length, area, or volume respectively in one-, two-, or three-dimensional space). The cumulative frequency curve for a_i against the proportion of observations $F = i/N$ lying within $a_{(i)}$ of observation P may then be plotted (Fig. 4.11). The ideal cumulative distribution function is also plotted. Both lines will start at zero and a y co-ordinate representing a proportion $F^* = F = 1/N$ of the population (not proportion zero since observation P itself lies at a distance of zero). The 'ideal' distribution F will plot as a straight line terminating at $a = A$ (the total length, area, or volume within which sampling was performed), and proportion $N/N = 1$. The empirical cumulative distribution will not generally reach proportion N/N until some value A_1 higher than A, because of geometrical effects. Consider the case of an observation at the corner of a cubic sampling space with sides of length one. The distance from this observation to the most distant could be as much as $\sqrt{3}$, and the spherical volume which corresponds to this distance (as radius) is $4/3\ \pi(3^{3/2})$, or approximately 21·8, significantly greater than the volume $A = 1$ of the sampled space. However, this effect is usually not important

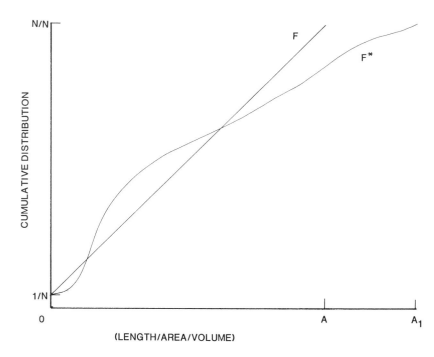

FIG. 4.11.   Ideal (F) and observed (F^*) distribution functions for the distances between one observation and every other observation.

as it is the zone close to $a = 0$ that is significant, because this is where *positive* deviations from ideal sampling density are most likely to occur. The maximum positive deviation is obtained as the one sided Kolmogorov statistic D^+ which is computed from the ideal distribution function $F(a)$ and the empirical distribution function $F^*(a)$ by

$$D^+ = \sup \, (F^*(a) - F(a))$$

or better still, since it is the *ratio* of the ideal to the observed sampling density which is of interest

$$D^+ = \sup(F^*(a)/F(a))$$

This latter equation is a little different from the standard Kolmogorov formulation, but may be transformed into it quite readily by using logarithms, so that an exactly equivalent expression is

$$\log D^+ = \sup (\log F^*(a) - \log F(a))$$

Lewis (1977, p. 220) has suggested the use of a method very similar to this in a geographic context, though he erroneously uses a straight line for an ideal cumulative distribution of distance in what is clearly a two-dimensional case — this demonstrates quite clearly the need for awareness of the geometric properties of the data before commencing analysis.

The D^+ value generated by the above method is only an intermediate step. Instead, however, of going on to use standard test procedures, the values known at this stage can be used directly to give an estimated $w_c(P)$ value. If $F^*_{D^+}$ and F_{D^+} are the empirical and ideal distribution function values respectively, at the distance at which D^+ occurs — the position of maximum positive deviation — then the weighting can be defined as

$$w_c(P) = F_{D^+}/F^*_{D^+}$$

or the ratio of theoretical to observed sampling densities at the distance from P where the sampling density around P is at a maximum. If there is no positive deviation from the ideal line, then this maximum occurs at $a = 1$, $F(a) = F^*(a) = 1/N$ and the ratio is thus one.

If ten observations are all taken at the same point, it is likely that they will provide the maximum positive deviation, at $a = 0$, $F(a) = 1/N$, but $F^*(a) = 10/N$; hence $D^+ = 10$ and $w_c(P) = F(a)/F^*(a) = 0.1$. This is the value which was defined intuitively at the beginning of this section.

At the other extreme, if there are very few observations close to P, then D^+ could refer to a maximum positive deviation at some large value of a; in this case, however, $F(a)/F^*(a)$ will tend to a value of one whatever the absolute magnitude of the difference between F and F^*. The larger the value of a, the more observation points it takes to move the value of this ratio away from one. It is possible to create more refined measures of clustering (for example, the Sherman (1950) statistic might be used); but it is recognised that the Kolmogorov type of measure provides a stable and empirically useful estimate of the required cluster weighting, just as an inverse power distance weighting is stable and useful. Both share the property that they are simple to compute. The cluster weights have one great advantage over the kriging weights they are intended to replace: that they need to be computed only once for the whole data set, and not separately at each point to be estimated.

4.3.2.4 Trend

It is very often possible to isolate a component of the surface which can be accounted for by a single global drift or trend of simple mathematical form. There is no reason in nonparametric geostatistics why this trend should not be removed from the data points before starting the interpolation procedure and added in to the estimated surface. Indeed, two results of doing this would be beneficial: in most cases the range of error in the estimated surface would be reduced, and a limited degree of extrapolation beyond the sampled area would be possible.

Because of the simplicity of the assumptions required by nonparametric geostatistics, there is no loss of validity of the estimated surface through the subtraction of this trend — in contrast with parametric geostatistics, in which great care must be taken when removing regional drift since the variogram is sensitive to changes in the trend component of the surface. Although universal kriging provides a limited solution to this difficulty, there is no general solution available for this problem in parametric geostatistics.

In nonparametric geostatistics the procedure which might be adopted is illustrated in Fig. 4.12 for a one-dimensional data set. It can be seen that any smooth surface, whether or not it has a simple mathematical expression, may be treated as a trend for this purpose, since none of the assumptions of nonparametric geostatistics are violated: the addition of two smooth surfaces produces another smooth surface.

4.3.2.5 Computation of a Weighted Varying Quantile Estimator

Having defined the distance and cluster weighting factors, it is now possible to recompute an estimate of q which takes into account the different contributions, in terms of information, made by the

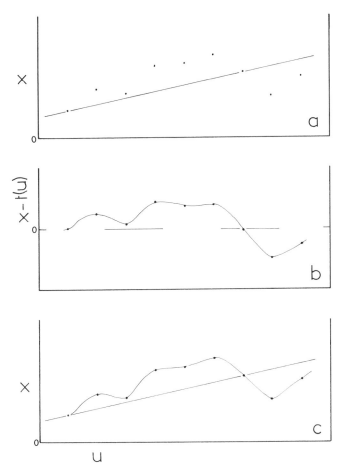

FIG. 4.12. The removal of a trend. (a) A smooth function is defined objectively or subjectively to describe a general trend or other large-scale features of the data. (b) The values of this trend computed at data points are subtracted from the data, and nonparametric geostatistics used to fit the residuals. (c) The trend is added to the fitted curve to obtain the complete fitted function.

observations, both in absolute terms as controlled by the geometry of the sampling pattern, and relative to the position of the point at which the estimate is being made.

A simple modification of the Mann/Kendall method would be to include weights in the computation of each comparison of x_i

with x_j. Since each comparison individually provides an unbiased estimate of τ, any weighting function which is independent of x_i and x_j may be used. The resulting modified tau will remain unbiased since it is a linear combination of a set of unbiased estimates. Thus, instead of recording $+ 1$ for $x_j > x_i$ and $- 1$ for $x_j < x_i$ (with $d_j > d_i$), an indicator which could take any value between $- 1$ and $+ 1$ might be computed. A simple such indicator might be

$$I = \pm\ w(i)w(j)$$

where $w(i)$ is the resultant (product) weighting on observation i, representing an assigned probability estimate that observation i is independently a member of the (fuzzy) neighbourhood set of P: it is computed by multiplying the distance and cluster weightings and any other weighting which might also be deemed necessary in particular cases. The indicator I takes positive or negative sign depending on the direction of decreasing value of x. The denominator of the expression for tau is no longer $\frac{1}{2}n(n - 1)$, or the total number of comparisons made, but must now be the total possible sum of weighting functions:

$$D = \sum_{i = 1}^{n - 1} \sum_{j = i + 1}^{n} w(i)w(j)$$

and the modified tau value will be

$$\tau^* = \sum_{i = 1}^{n - 1} \sum_{j = i + 1}^{n} I_{ij}/D$$

with q estimated by

$$\hat{q} = \tfrac{1}{2} + \tfrac{1}{2}\tau^*$$

This estimate of q may be applied to the weighted observation values to obtain the qth weighted quantile — exactly analogous to the weighted median introduced in Section 4.3.1, which is the 0.5th weighted quantile. An indicator function I as simple as that defined above would have practical difficulties, in the form of the estimated surface, which would have discontinuities wherever observation distances are equal: consider, for example, the case of a point P which is equidistant from two observations i and j. A small spatial displacement in one direction will produce a finite weighting $w(i)w(j)$ which adds to the tau value. A small displacement the other way produces the same finite weighting subtracted from the tau value. The problem arises mainly when observations i and j are on opposite sides of P. Intuitively, there seems little justification for using comparisons between such pairs of observations as estimators of q in any case; it would be desirable to include in the definition of I two additional components: a smoothing function which would prevent the appearance of discontinuities in the estimating surface — at least arising from the relative positioning of points i, j, and P. It would also be useful to include an angular function which depresses the weightings of comparisons between observations which are on opposite sides of P. Such a function would be $\cos\frac{1}{2}\alpha_{ij}$, where α_{ij} is the angle subtended at P by the two observations i and j. A smoothing function which could be used might be based on the relative distances d_i and d_j; the function

$$s = (d_j - d_i)/(d_j + d_i)$$

would tend to zero as the two observations approached equal distance from P. The indicator I might thus be redefined as

$$I = \pm w(i)w(j) \; s \; \cos\tfrac{1}{2}\alpha_{ij} \quad \circ$$

This remains a valid function for use in the estimation of tau, and of q, and the estimate remains unbiased because I is still defined independently of x_i and x_j. Note that with any redefinition of I, D must also be redefined in exactly the same way.

The effect of such a redefinition is to eliminate comparisons between observations which are exactly on opposite sides of P — in one dimension this prevents the simultaneous use in comparisons of points to the left and to the right of P, since such pairs automatically subtend an angle $\alpha = \pi$ at P.

Having computed an estimate of q, it remains to determine the quantile *value*. For this purpose, the observations are weighted by the simple weightings $w(i)$ which are, as defined above, merely the product of distance and cluster weightings, and any special weightings included for specific purposes.

The weighted varying quantile estimator is the basis upon which a practical nonparametric estimation technique for spatial data may be built; it provides inherent flexibility, in that weighting schemes may be changed and improved upon without the necessity to alter the method of estimating q and deriving the quantile. Given an unbiased estimate of q, the varying quantile estimator is itself an unbiased estimator of the true value at any point P, which the median is not, except in a few special cases. However, it does have some practical drawbacks not shared by the weighted moving median, and discussed in Section 4.5.2, in which a hybrid technique is proposed, sharing some of the merits of each.

4.4 ESTIMATION OF ERROR

Although the computation of the error of estimation is not a fundamental part of the method of nonparametric geostatistics, it is not too difficult to estimate. In parametric geostatistics, a formalised error estimate is generated which assumes that errors are normally distributed with zero mean: in nonparametric geostatistics, if required, not only an estimate of error magnitudes but also an indication of the form of error distribution may be obtained by use of the jack-knife method (Tukey, 1970), in which the estimate of q is made n times with one observation removed from the data set each time. From these n estimates each based on $n - 1$ observations, the distribution of $\hat{q}$ is obtained, and it is an easy matter to derive upper and lower confidence bounds based on the available sample. These upper and lower limiting values of q are then applied to the data to derive the upper and lower x value bounds. It should be emphasised that the error determined in this way will generally underestimate the *real* error which relates the observations to the true surface — just as the parametric geostatistical estimates of error tend to underestimate the real error.

The discrepancy between the error determined by the jack-knife procedure and the real error will depend on how close the observations are to P, in comparison with the complexity of the

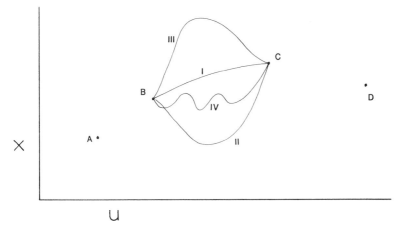

FIG. 4.13. It is always possible to interpolate an uncountably infinite number of different smooth surfaces between two point observations B and C. In the presence of other data points such as A and D, some of the curves (e.g. I) seem more likely than others (e.g. II, IV).

real surface form. It is possible to determine one of these factors at every point to be estimated, since d_{min} is the distance from P to the closest observation. This allows *relative* estimation of the discrepancy, provided that the surface complexity does not change from one place to another. This assumption brings in another type of stationarity, however, and may not hold true in the majority of cases; moreover, the relationship between d_{min} and the surface complexity cannot readily be defined — indeed the complexity of the real surface is unknown and unknowable, since however close together point observations may be made it is always possible to interpolate an uncountable infinite number of different smooth surfaces between them (Fig. 4.13). Value and error estimation are based on the assumption that in the presence of all the surrounding data points, some surfaces are more likely than others.

4.5 SOME PRACTICAL PROBLEMS

In real sampling programmes, there are a few departures from the idealised model which could cause difficulties in interpretation and use of nonparametric geostatistics. Some of these problems are shared by parametric geostatistics; some are less severe for nonparametric than for parametric methods.

4.5.1 Anisotropy

If the true value varies more quickly in one direction than in another — like the surface form of a hillside — it displays geometric anisotropy. An alternative expression of geometric anisotropy might be a succession of linear ridges, in which the degree of variation to be expected at right angles to the crest line is much greater than along it.

In standard parametric geostatistics, such anisotropy must be allowed for specifically, by adjustment of coordinates to create an illusory isotropic data set, or by using different variograms for each different direction between the point to be estimated and the sample points. In nonparametric geostatistics, geometric anisotropy is an accepted property of the data, and needs no special treatment: it is not a theoretical problem. However, if the form of the anisotropy is approximately known before analysis, it can be of great benefit to remove it as a deterministic component of the surface, and reduce the range of errors in the estimated surface, which will now consist only of the 'residuals'. This procedure would be of highly doubtful validity in parametric geostatistics, since removal of a trend alters the form of the variograms and can produce gross departures from the optimal estimation which kriging is intended to provide. The only parametric geostatistical methods to approach this problem are Universal Kriging (UK) which incorporates the equation for a polynomial trend surface within the kriging equation system, and more advanced kriging methods which have been developed from UK. All are restricted by the range of surfaces which can be expressed in the form of a finite power series; no such limitation applies to nonparametric geostatistics, which demands only a smooth surface. It is quite permissible, indeed, to use one estimated surface to approximate the trend when computing another (as may be required, for instance, to interpolate both the top and bottom of a coal seam).

A second form of anisotropy is not so easily handled: it is expressed as qualitative rather than quantitative variation, or a separation of the data set into zones. For this reason it is commonly known as zonal anisotropy. An example of this is presented by a sequence of sediments in which beds of quite different rock are separated by relatively sharp boundaries. In both parametric and nonparametric geostatistics, these sharp boundaries pose insurmountable difficulties, since the assumption of continuity is violated.

4.5.2 Extraction of Quantile q from the Data

In Section 4.3.2 it was noted that although the varying quantile method provides a better point value estimate in any number of dimensions than does the weighted median — because of the geometric problem — this does not necessarily imply that the estimated surface will be smooth. Consider the case of a point P near to an observation x_j which is higher in value than surrounding observations. Assume, in fact, that P lies closer to j than to any other observation. Since x_j is greater than the other nearby observations, there will be a high correlation τ between distance and downward slope, and the q value will be high. However, there is little need to know the q value at points close to an observation, since the observation x_j itself is a good estimator. Using the varying quantile method, a high value of q might well suggest that the true point value at P is best estimated by some value between x_j and infinity. As there is no information available on the form of the distribution for values greater than x_j, the best varying quantile estimate for x_p must be taken to be x_j. The effect of this will be that if the varying quantile method is used, any peaks and troughs will tend to be surrounded by flat areas in which the estimated value is identical to the peak or trough value. This effect, in terms of the estimation procedure and of the estimated surface, is illustrated in Figure 4.14.

There is an empirical way to avoid this effect to some extent, if, as in many applications, it is considered undesirable. It may be rationalised as follows: very close to an observation, the distance weight on that observation will be unity, but the weights on all

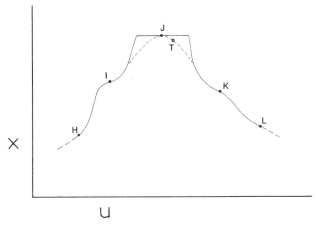

FIG. 4.14. The varying quantile method (solid line) tends to produce estimated plateaux around extreme values (J). The proposed hybrid method produces a smooth estimated curve in the area near to extreme values (broken line).

other observations will be small (using the distance weighting $w_d = (d_{min}/d_i)^k$); the sum of the weights is used to create a hybrid estimator q':

$$q' = \tfrac{1}{2} + \frac{\tau}{2} \left(\sum w_d(i) - 1\right)/\sum w_d(i)$$

For points p close to an observation j, q' will be close to $\tfrac{1}{2}$ since the sum of weights is close to one. Thus the weighted median is used at point J in Fig. 4.14, as it is at points H, I, K and L. However, at positions of P intermediate between observations (as at F in Fig. 4.14) the distance weight on the closest observation will still be one, but weights on other observations will *not* be negligible since the ratios d_{min}/d_i will be higher. Thus τ' as calculated above will be allowed to deviate more from a value of $\tfrac{1}{2}$. The result is that away from observations, the value of q' will be little changed from that predicted by use of the modified Mann/Kendall statistic, but in regions close to observations the value of q' to be used will lie between that predicted by τ and $0·5$ (the median). This adjustment will therefore produce a surface having the stability and smoothness which would be obtained from the weighted moving median method. The surface would lie close to the weighted median in areas near observations, but would approximate to the pure varying quantile surface in regions further from the data points.

Although such a hybrid approach does have practical attractions, it nevertheless lacks the theoretical basis which is a property of either the weighted median or the varying quantile approaches.

It is noteworthy, however, that *at* the locations of data points, particularly if several observations are coincident, the median is usually the appropriate statistic to use as estimator, rather than some undefined quantile.

4.5.3 Finite Support of Observations

In the purely observational sciences such as astronomy, meteorology, and some branches of geophysics, observations can realistically be considered as point values, and are assumed to be repeatable: it is possible to make any number of observations at one spatial location (though only one if time is included as a coordinate). In contrast, most geological and mining studies are concerned with physical rock samples, with an observation corresponding to a piece or pieces of rock of finite volume. It is obvious that only one such piece of rock can be collected from a particular location, since after collection that location is left empty.

The observation usually consists of an assay or geochemical analysis, or some measurement of a physical property of the piece of rock, and is intended to estimate the *average* value of a given parameter in that rock. The shape and volume of the rock are collectively termed the 'support' of the observation. If the dimensions of the support are very small in comparison with the sampling area or volume (e.g. centimetre sized rock samples in a kilometre sized area), the problem of finite support may be ignored and the observations treated as point observations. In many mining applications, however, the support dimensions are not negligible. For example, in a set of drill cores, the total length of core in each hole may be of the same order as the total depth of the ore-body. If the core is divided into a number of lengths for analysis, each observation will relate to a cylindrical support of significant length. Treatment of these observations as point values would introduce a smoothing error which could be quite significant. This smoothing error could be minimised, in the case of linear cores, by first using nonparametric geostatistics to interpolate at frequent intervals along the core. The interpolated points could then be used in place of the original observations. They would, of course, require to be weighted as an indication of the number of observations from which they were computed: if k observations along a core are replaced by K interpolated points, the appropriate weighting factor for each of these points is k/K. This weighting is extra to those applied for clustering and distance, and is multiplied by those to obtain the resultant weighting on each point.

4.5.4 Estimation of Finite Volumes

One of the principal purposes to which parametric geostatistical methods are put is the estimation of average values within blocks of finite volume. These methods are capable of direct application to such problems by adjustment of the system of kriging equations provided that the blocks are of simple shape. Thus for spherical, cubic, or cuboid blocks, there is no difficulty; for other geometric shapes or for irregular shapes, however, there is no such simple solution. Indeed, even for the regular shaped blocks, the simple solutions available are in fact only simplifications of the more general case. The problem may be stated very simply: to estimate the following integral

$$\bar{X} = \int^v x/V$$

where x is the value at a point within the block volume, V is the volume, and $\bar{X}$ is the block average. Now, in the general case, the accuracy of $\bar{X}$ depends on the accuracy of each individual $\hat{x}$ estimate. If an estimation method is chosen which minimises the variance of errors in $\hat{x}$, then the variance of $\bar{X}$ will also be minimised. This is the approach taken by parametric geostatistics. An alternative approach would be to minimise the absolute error in each $\hat{x}$ estimate; the result is an $\bar{X}$ estimate as close as possible to the true $\bar{X}$. In either case, the method of computation of $\bar{X}$ is the same — a net of points is set up, covering the volume uniformly, and $\hat{x}$ is computed at each point. The simple arithmetic average of these values is then the desired estimate of $\bar{X}$. It should be noted that an average is used here (and not a median) even in the case of nonparametric geostatistics because the mathematical operation being simulated by this discrete operation is that of integration, whose discrete equivalent is *summation*.

Because of the significance of applications which require the use of block averages, this method will be described in more detail in the following chapter.

4.6 SUMMARY: PROPOSED ESTIMATION METHODS

Spatial interpolation methods such as moving average, kriging, moving median, or varying quantile, may all be considered as two-stage processes. The first stage consists of an estimation of the local distribution function using some set of weightings on the observations to reflect an assumed probability of their inclusion in a local neighbourhood set. The second stage is the estimation from this distribution of the appropriate 'central' value which best represents the unknown value at the point to be interpolated. Weighted average and most kriging methods use the mean of this local distribution; the moving median uses the median; the varying quantile uses a quantile whose position is predicted by an assessment of the local surface form. In all cases, the 'local neighbourhood' is defined as a fuzzy set, with the probability of independent membership being assigned by some standardised procedure of computation of weighting factors; in kriging, for example, this procedure consists of the solution of a set of n simultaneous equations. For the nonparametric techniques introduced in this volume, a simpler procedure has been proposed — although there is no reason why kriging weights should not be used when the assumptions required by the kriging techniques are satisfied (especially the requirement that there be a sufficient degree of stationarity to allow the use of the same variogram over the whole region of interest). However, for the second stage, that of determining the 'centre' of the estimated local distribution, the mean is discarded in favour of either the median — which produces a robust estimate guaranteeing minimum absolute deviations rather than minimum variance — or a freely varying quantile whose value is determined by the use of a nonparametric correlation measure (a modified Kendall τ) to estimate the 'peakness' or 'troughness' of the surface in the local neighbourhood.

Although the varying quantile method should give estimates closer to the 'true value' than either a mean or median, it can lead to practical problems, since it is less stable than the median (and certainly less robust) and leads to estimation problems near major peaks and troughs. A hybrid method has been suggested to overcome these problems by forcing the quantile to tend toward the median at points close to observations, but such a measure does not have the theoretical justification of either the median or the varying quantile.

For practical purposes the most stable of the estimators proposed in this chapter is the weighted moving median, though intuitively better interpolated surfaces are obtained using the hybrid method. The median has the statistically 'nice' property that it is the minimum absolute deviation estimator of the centre of the local distribution, and this property can be exploited in some practical applications as will be seen in the next chapter.

CHAPTER 5

Mining Applications

5.1 BLOCK VALUE ESTIMATION

The majority of estimation problems in mining (and in some other fields) are concerned not with point values but with average values over regions of space (or blocks) of finite size. The value to be estimated is thus the *integral*, over the volume of the block, of the values at all points in the block.

It is common practice by now to use one of the many kriging methods to solve almost any problem of this sort; provided that the assumptions are satisfied, a kriging method will generally produce the minimum variance estimate. However, this estimate *is* sensitive to departures from the ideal case. The weighted median provides an alternative to the kriged weighted average for point estimation, providing a minimum absolute deviation unbiased estimator which is stable with respect to distribution. The disadvantage associated with use of the weighted median is that it does not yet have a supporting body of statistical theory (such as regionalised variable theory), and it is not possible directly to extend the point estimation method for block estimation as can be done in kriging methods. However the weighted median estimator can be used indirectly, by approximating a block value with the

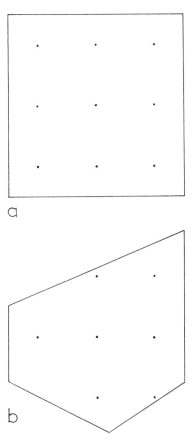

FIG. 5.1. Estimation of a regular (a) or an irregular (b) block from a set of
point estimates uniformly distributed within it.

average of a discrete set of point estimates evenly distributed
within the block (Fig. 5.1). This is similar to the procedure which
must be used in kriging to estimate the value of an irregularly
shaped block.

At each point where an estimate is to be computed, the estimation error may be expressed as a deviation

$$\varepsilon_p = \hat{x}_p - x_p$$

and the absolute value of this error is minimised by using the weighted median as point estimator. The median is necessarily also unbiased (it bisects the distribution); thus if a number of medians are summed, the corresponding sum of deviations tends to zero. The block estimation error may be defined as the median of the *absolute values* of these deviations. There is a difficulty, however. It is already known that the median is *not* in general the best estimator of the true value, since the varying quantile takes into account the local surface form as estimated from the data — additional information that is not used in computing the median. Moreover, the weighting scheme used may not be optimal in the sense that the distribution represented by the set of weighted observations may not be representative of the true distribution in the local neighbourhood of P. On a large scale, these difficulties may not be relevant, but at the scale of estimation of each block they are significant, and result in a gratuitous degree of smoothing. As in some kriging methods, the weighted median will tend to underestimate 'peak' block values and overestimate 'trough' block values.

If the surface form is taken into consideration (and assuming that the weighting scheme *does* provide a reasonable estimate of the local distribution), the varying quantile is a usable unbiased estimate of the true point value. Although a model surface generated using point estimates computed by this method may be discontinuous — as noted in Section 4.5.2 — these estimates may be used quite validly to produce estimated block values which will reflect more accurately the local distribution.

Whichever estimate is used, the algebraic sum of errors ε_p will necessarily always be less than the sum of absolute values of errors, when taken over the region of interest. Thus the average of the $\hat{x}_p$ estimates is taken as the block value estimate, but the block estimation error is *less* than the average point estimation error. Indeed, the expected value $E(\varepsilon_V)$ of the estimation error ε_V for block V must be zero if the point estimates are unbiased. In terms of assessing the reliability of the block estimate, therefore, it is the uncertainty of ε_V which is of significance. The distribution of ε_V depends entirely on the distribution of the true values, however, since at any one point ε_p is defined as the difference between two variables $\hat{x}_p$ and x_p, with different distributions: $\hat{x}_p$ (if a median) is asymptotically normally distributed, but x_p (the true value) has a point distribution since it is a constant (albeit an unknown one) which is defined solely by the geological processes. Given a sufficiently large number of estimates $\hat{x}_p$ with associated error ε_p, the distribution of ε_V will be approximately normal, since it is the sum of a large number of continuously distributed variables.

For this reasoning to apply it is required that the estimated $\hat{x}_p$ be unbiased: hence the weighted median may be less appropriate than the varying quantile over any small part of the population.

In practice, it is not possible to test one against the other to determine which is better suited to the computation of block estimates, since the true values x_p are generally unknown, as is the true block average value x_V. However, the deviation ε_p may be *estimated* for each point estimate by computing the weighted average of differences between the estimated value and each (weighted) observation value used in its generation. What cannot be estimated using the median alone is the *sign* of this deviation. All that can be expected is that roughly half will be positive and half negative. If, therefore, the estimated error terms ε_p over a block are sorted into ascending order, and then alternate terms arbitrarily allotted a negative sign, the mean of the resulting set of numbers will be close to zero in general, and the standard deviation will constitute a measure of the block estimation error.

If the point estimates are computed by the varying quantile method, the mean deviation is more difficult to compute; consider the case of a predicted quantile q greater than $\frac{1}{2}$. If the ordinary weights (as used for the weighted median) are used here, there will be an excess contribution to the error value which is to be estimated, made by observations with low x_i values. The purpose of the varying quantile method is to compensate for the effect of surface forms which produce such an excess (for example peaks and ridges in this case) by taking the median of a progressively smaller portion at one end or other of the distribution. Hence a mean deviation could perhaps be estimated by computing the weighted average of deviations within the largest symmetric zone which may be placed around quantile q: from quantile 0 to $2q$ for $q < \frac{1}{2}$ or from quantile $1-2q$ to 1 for $q > \frac{1}{2}$.

There is a special problem when the estimated q is one or zero, since in such cases the *estimated* error is zero but the *likely* error is in fact greater than at most other points. There is no clear solution to this problem if the pure varying quantile is used, but the hybrid varying quantile/weighted median technique suggested in Section 4.5.2 would resolve the difficulty and would also neatly express the increased uncertainty in the zones close to peaks and troughs.

5.1.1 Selective Mining, Cut-off Values, and Detection Limits

Estimated block values and errors are of use in this simple form only when the smallest mining unit is at least as large as the computed block. In many cases it is possible, however, to selectively mine a part of a block. To allow proper estimation of a selectively mined block, and to predict correctly the mined volume and grade, it is necessary to estimate the distribution within the block.

The within-block distribution is assumed by linear kriging to be a simple normal distribution: if parametric geostatistics are to be used, then it requires either a disjunctive kriging approach to provide more detailed estimation of this distribution, or linear kriging of a number of subdivisions of the block which are then cumulated to provide the overall block statistics. The latter approach is adopted by nonparametric geostatistics — a set of point estimates $\hat{x}_p$ within the block is cumulated.

If a part of the block consists of waste, and another part consists of ore, then selective mining will ideally remove all of the ore and none of the waste — or at least will identify the two and treat them separately. The classification is usually made on the basis of one or more 'cut-off' values — ore grade, impurity content, or physical properties being typical constraints. If a proportion r of estimated point values $\hat{x}_p$ within the block lie between cut-off limits x_L and x_U (being the lower and upper limits respectively of the range of acceptable values) then the associated selectively mined volume and tonnage may be appropriately adjusted downward and the mineable grade estimated by the mean of values which lie within the limits.

Thus far, nonparametric geostatistics provides capabilities similar to those available from parametric geostatistics. However, if some of the recorded values of observations are 'below the detection limit' of the analytical technique used, then parametric geostatistics cannot validly be used (Fig. 5.2). The reason is that such observations are not allocated a numeric value — merely a statement that if it could be recorded it would lie somewhere between zero and the detection limit. There is as yet no theoretically justified way to handle such data with parametric statistical methods. Using nonparametric statistics, however, there is no problem, since such values are sorted to the lower end of the distribution. If a weighted median is used, and over half of the distribution consists of such values, then the estimated $\hat{x}_p$ is given the same value (probably in practice some conventional code is adopted, to be interpreted as 'not detected'); similarly, if the varying quantile or hybrid methods are used, then the estimated $\hat{x}_p$ is recorded as 'not detected' if at least the lower q proportion of the distribution is recorded as such.

In the selective mining process, a value below a detection limit

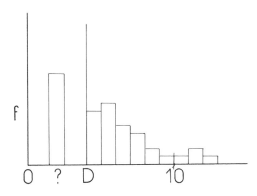

FIG. 5.2. A common problem in geochemistry: measurements cannot be made below a detection limit D. Observations in which the true value is below D cannot be recorded at any particular value on the histogram — the true form of the distribution for values below D is unknown.

may generally be treated as a zero value since a cut-off limit is of no practical use if it lies within a region of values inaccessible to the analytical instruments available. Thus there is usually no practical problem caused by the presence of such non-numeric values, apart from that of handling them with parametric statistical methods.

5.2 RECONCILIATION OF ESTIMATED AND ACTUAL VALUES

The quality of an estimation method is determined by the consistency and reliability of the estimates it produces. It is sometimes possible to assess the estimation method in use at a mine by comparison of the originally estimated, predicted values, with the actual mined values of the same blocks. This comparison of estimated and actual values may be illustrated in the form of a simple scatter plot (Fig. 5.3). The general features of such a plot may be considered schematically as in Fig. 5.4, in which the scatter of points is replaced by its bounding line. The two straight lines, horizontal and vertical, represent the cut-off grade, and divide the scatter into four regions. In region I, actual values are above the cut-off and are estimated as such. In region II, actual values are below cut-off as predicted, and are correctly treated as waste. Regions III and IV contain blocks for which incorrect decisions are made on the basis of estimates. In region III, blocks estimated as waste are in fact economically mineable. In region IV, blocks estimated above the cut-off are in fact waste. An important objective of any block estimation technique, therefore, is to minimise the number of points which plot in regions III and IV.

In an ideal unbiased estimation method, the numbers of points which lie within these two regions will be equal. This is the intention of kriging methods: to provide an estimator which *on average* will produce estimated values lying along the 45° line ($x_V = \hat{x}_V$). However, there is a problem even when such an aim is achieved: the costs of region III errors and region IV errors are different. It is evident that in most cases the cost of mining and milling waste

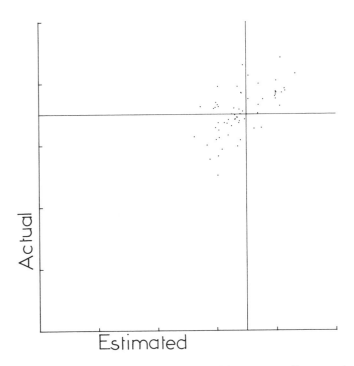

FIG. 5.3   Comparison of estimated values with corresponding actual values by means of a scatter plot. A cut-off grade (at 35) for economic extraction is indicated.

will be very different to the cost of leaving ore untouched or simply moving it to a waste dump or stockpile. A region III error is usually less costly than a region IV error. An ideal estimation method therefore is as *consistent* as possible, but will not be unbiased. The scatter of points should lie as close as possible to the 45° line, but any errors which push points away from this line should move them preferentially into region III rather than region IV. The optimum proportions of points in the two regions will of course be a function of the relative costs of such errors.

Compared with the kriging estimators, the weighted median provides conservative estimates. In a positively skewed distribution such as occurs in many mining situations, the magnitude of overestimates will generally be less than that of underestimates

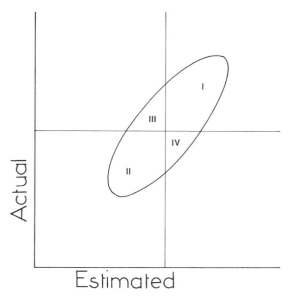

FIG. 5.4. A schematic generalised version of Fig. 5.3 showing the cut-off grade, dividing the cloud of points into four separate classes. Areas III and IV contain observations which have been misclassified because of error in the estimated values. If the data are used for decision making, a type IV error tends to be more costly than a type III error because the block is incorrectly estimated as being worth mining.

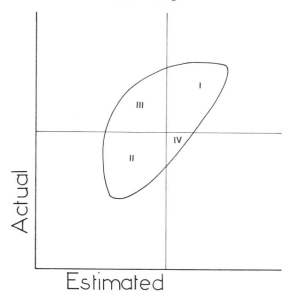

FIG. 5.5. Using nonparametric (weighted median) estimates, serious errors tend to lie in area III rather than in area IV.

and the fact that a block estimate is an average of a number of weighted median point estimates does not materially alter this. Therefore serious errors will tend to lie on the region III side of the 45° line rather than on the region IV side (Fig. 5.5).

5.3 A SIMULATED IRON ORE DEPOSIT

Clark (1979) devised a synthetic data set to demonstrate the block kriging method. It (Fig. 5.6) consists of a 400 m × 400 m square area with 50 random samples within it. The locations and grades of these samples are listed in Table 5.1. These data were used to produce linear block-kriging estimates of 64 square blocks 50 m × 50 m in size (Fig. 5.7). The kriged estimates may be compared with the simulated 'true' values (Fig. 5.8) to produce the deviations $x_p - \hat{x}_p$ (plotted in Figs. 5.9 and 5.10). The sum of squared deviations and the sum of absolute values of deviations are listed in Table 5.2, together with the algebraic sum of deviations which

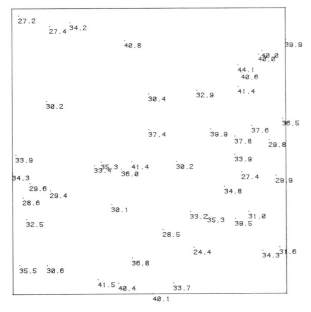

FIG. 5.6.   Grade data for the simulated iron ore deposit, listed in Table 5.1.

TABLE 5.1
Location and grade data for simulated iron ore deposit
(reproduced with permission from Clark, 1979).

Easting	Northing	%Fe	Easting	Northing	%Fe
0	170	34·3	5	195	33·9
10	40	35·5	20	105	32·5
15	135	28·6	25	155	29·6
55	145	29·4	50	40	30·6
125	20	41·5	155	15	40·4
175	50	36·8	145	125	30·1
120	180	33·4	130	185	35·3
160	175	36·0	175	185	41·4
240	185	30·2	220	90	28·5
260	115	33·2	205	0	40·1
235	15	33·7	265	65	24·4
365	60	34·3	390	65	31·6
285	110	35·3	325	105	39·5
345	115	31·0	310	150	34·8
335	170	27·4	385	165	29·9
325	195	33·9	325	220	37·8
350	235	37·6	375	215	29·8
290	230	39·9	200	230	37·4
10	390	27·2	55	375	27·4
85	380	34·2	395	245	36·5
50	270	30·2	165	355	40·8
200	280	30·4	270	285	32·9
400	355	39·9	365	340	40·0
360	335	40·0	330	320	44·1
335	310	40·6	330	290	41·4

may be considered as an indication of the degree of bias (since an unbiased estimate will have a zero algebraic sum of deviations). A histogram of these deviations (Fig. 5.11) shows a fairly sharp peak, and rather long tails: this is a typical distribution of kriging errors, and close to that predicted by regionalised variable theory.

For nonparametric geostatistical estimation of block values from the same data set, it is necessary first to estimate the weightings for each observation and for each point to be interpolated.

27.5	31.7	36.9	38.0	37.9	37.9	40.1	40.5
29.0	31.7	35.9	36.5	35.2	37.3	41.4	39.8
30.5	31.7	33.8	33.6	32.1	35.2	40.0	38.3
32.4	32.0	34.1	37.9	34.9	36.8	37.6	33.5
31.7	31.3	33.6	37.1	32.7	33.0	31.6	29.4
30.3	31.1	31.0	31.8	31.2	34.4	34.5	30.8
33.3	32.7	34.9	33.6	29.0	29.1	35.2	32.5
33.4	33.7	39.4	38.8	33.8	30.2	32.9	32.2

FIG. 5.7.   Estimates for 50m × 50m blocks, produced by linear block kriging.

36.2	35.0	43.0	44.2	37.5	38.5	40.5	38.7
36.1	28.2	38.4	36.3	30.8	35.2	41.6	38.5
39.5	25.1	34.2	33.5	29.3	36.0	39.2	32.8
39.5	31.2	38.5	40.0	36.5	38.2	37.4	32.8
33.1	32.4	34.5	38.5	36.2	33.9	30.0	30.9
35.9	36.9	36.5	33.6	34.9	34.5	34.2	33.6
35.0	34.1	37.5	32.7	27.7	31.4	35.5	34.2
41.3	31.6	38.8	37.9	33.6	29.8	35.8	35.0

FIG. 5.8.   'True' block average values for the simulated iron ore deposit.

8.7	3.3	6.1	6.2	-0.4	0.6	0.4	-1.8
7.1	-3.5	2.5	-0.2	-4.4	-2.1	0.2	-1.3
9.0	-6.6	0.4	-0.1	-2.8	0.8	-0.8	-5.5
7.1	-0.8	4.4	2.1	1.6	1.4	-0.2	-0.7
1.4	1.1	0.9	1.4	3.5	0.9	-1.6	1.5
5.6	5.8	5.5	1.8	3.7	0.1	-0.3	2.8
1.7	1.4	2.6	-0.9	-1.3	2.3	0.3	1.7
7.9	-2.1	-0.6	-0.9	-0.2	-0.4	2.9	2.8

FIG. 5.9.   Deviations of block values ('true' value minus kriged estimate).

As shown in Chapter 4, these weightings are computed in two parts, which are then multiplied together: the cluster weighting, computed once only for all data points; and the distance weighting, computed for each data point at each point to be interpolated. The cluster weightings for the synthetic iron ore deposit have been computed by the method of Section 4.3.2.3 and are shown in Fig. 5.12.

Nonparametric estimates of three types have been calculated for this set of data: a median point estimate computed at block centres, a hybrid median/varying quantile estimate at block centres, and a block average of nine point median estimates computed at evenly spaced points within each block (i.e. at the centres of nine equal squares within each 50 m square). The sums of squared deviations, absolute deviations, and algebraic sums of deviations for all of these techniques are listed in Table 5.2. It is apparent that the point estimates, median or hybrid, are not particularly good at predicting block average values; the median (Fig. 5.13) performs rather better than the hybrid method (Fig.

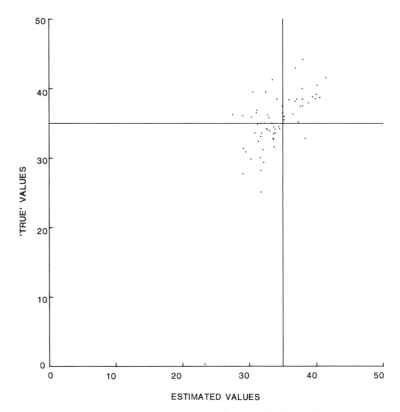

FIG. 5.10.   Scatter plot of 'true' block values against kriged estimates, with an
assumed cut-off grade of 35% iron.

5.14), but neither is as good as a block-kriging estimate. It is worth
noting, however, that the *bias* in both estimators (expressed as the
algebraic sum of deviations) is apparently less than the kriging
bias.

The nine point averaged block estimator fits the 'true' values
very much better than either of the cell-centre point estimators
used; the estimates it generates are shown in Fig. 5.15, and
deviations from the 'true' values in Figs. 5.16 and 5.17. As can
be seen from Table 5.2, its performance in general is comparable
with that of the block-kriging method — indeed the sum of squared
deviations is slightly lower than that of the block-kriging estimator. Also, the bias is appreciably less. Overall, therefore, the

TABLE 5.2

Deviations of estimated values from 'true' values in simulated iron ore deposit.

Type of estimator	Sum of squared deviations $\sum (x_p - x_p)^2$	Sum of absolute deviations $\sum |x_p - x_p|$	Algebraic sum of deviations $\sum (x_p - x_p)$
Kriging	755·32	161·00	82·00
Weighted median (point estimate)	828·75	180·56	75·98
Nonparametric hybrid point estimate	947·37	195·75	70·25
Weighted median (average of nine points per block)	754·55	167·04	71·49

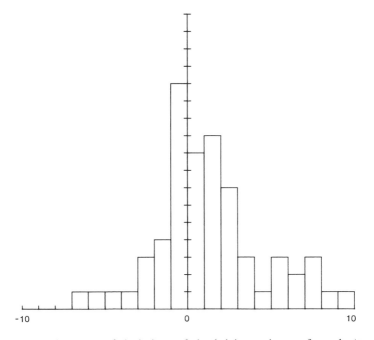

FIG. 5.11. Histogram of deviations of the kriging estimates from the 'true' values for the 64 blocks.

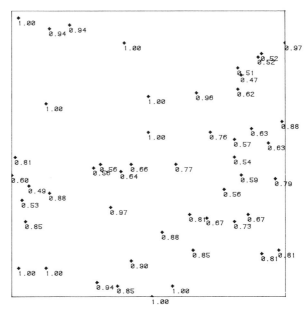

FIG 5.12. Nonparametric geostatistics: cluster weightings $w_c(i)$ computed for the 50 data points.

27.3	34.1	34.4	40.8	39.9	39.9	40.0	39.9
29.8	30.3	34.2	40.6	34.0	36.9	43.2	40.0
30.2	30.2	33.1	31.8	30.9	32.9	41.4	37.5
33.5	30.5	34.1	37.0	37.0	39.9	37.8	29.9
32.5	30.4	33.5	40.9	30.3	33.8	28.4	30.0
29.5	29.9	30.1	30.3	33.1	33.4	34.7	31.3
32.6	32.0	35.3	36.3	28.9	26.3	34.4	33.2
35.4	31.9	41.5	39.8	33.7	33.5	33.8	33.9

FIG. 5.13. Weighted medians computed at the block centres.

27.3	34.1	34.0	40.7	40.2	40.7	40.3	40.0
29.7	30.4	34.2	40.3	37.6	40.0	42.4	40.0
30.2	30.2	34.2	32.2	32.8	33.0	41.4	38.1
32.0	32.5	34.6	37.1	37.1	39.9	37.8	29.9
29.9	30.2	33.5	40.9	30.3	33.1	28.4	29.9
29.4	29.7	30.1	31.0	33.0	33.3	34.1	31.1
32.0	31.9	35.0	36.4	28.9	26.0	33.9	32.8
35.4	32.0	41.5	40.1	33.7	33.4	33.8	34.0

FIG. 5.14.   Hybrid method estimates computed at the block centres.

27.3	31.8	35.7	40.7	39.5	39.8	40.0	39.9
29.3	30.8	35.2	37.8	34.6	36.8	40.8	40.0
30.2	30.2	33.0	32.4	32.0	33.6	39.8	38.2
32.3	31.6	34.2	36.6	35.9	38.0	37.6	34.0
32.2	31.4	34.3	37.3	32.3	33.2	32.1	30.2
30.3	30.3	30.5	31.3	32.3	34.2	34.2	31.1
32.9	32.1	34.7	34.9	30.1	29.9	34.8	33.0
33.9	33.6	41.1	38.9	35.0	33.0	33.8	33.6

FIG. 5.15.   Weighted median block estimates computed by averaging nine weighted medians at uniformly spaced points (in the pattern shown in Fig. 5.1 (a)) within each block.

8.9	3.2	7.3	3.5	-2.0	-1.3	0.5	-1.2
6.8	-2.6	3.2	-1.5	-3.8	-1.6	0.8	-1.5
9.3	-5.1	1.2	1.1	-2.7	2.4	-0.6	-5.4
7.2	-0.4	4.3	3.4	0.6	0.2	-0.2	-1.2
0.9	1.0	0.2	1.2	3.9	0.7	-2.1	0.7
5.6	6.6	6.0	2.3	2.6	0.3	0.0	2.5
2.1	2.0	2.8	-2.2	-2.4	1.5	0.7	1.2
7.4	-2.0	-2.3	-1.0	-1.4	-3.2	2.0	1.4

FIG. 5.16. Deviations of block values ('true' value minus averaged weighted median estimate).

simple block-averaged weighted median is at least as good as the kriged estimate in this simple example which was designed as a demonstration of the kriging method.

The histogram of deviations of the nonparametric block estimator from the 'true' values (Fig. 5.18) shows a distribution with a less sharp peak and shorter tails than are evident in the corresponding histogram of kriging errors. Again, this is as would be expected from the theory, which suggests that there will be relatively fewer large deviations than produced by kriging.

This simple demonstration shows that in a case which one would expect to be ideal for kriging, the nonparametric block-averaging method can be as effective. No comparison has been made of the confidence bounds or estimation error, as these are computed upon entirely different bases for the two groups of methods. However, the observed errors — the deviations of estimates from the 'true' values — provide an objective measure of the relative performances of the techniques. It would be quite possible to

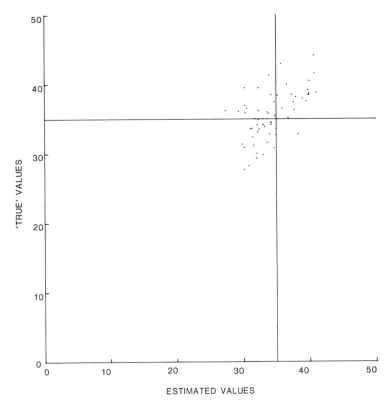

FIG. 5.17. Scatter plot of 'true' block values against averaged weighted median estimates, with an assumed cut-off of 35% iron.

increase the computing time and cost, and use nonparametric estimates based on larger numbers of point estimates per block (e.g. 16 or 25), or to average hybrid estimates rather than the median estimates used.

Although good estimates can be expected from the nonparametric geostatistical method when the conditions for validity of kriging are satisfied (such as the example used here, which was *constructed* to satisfy the kriging assumptions), the significant advantages of nonparametric geostatistics become apparent when these assumptions are not satisfied, and there are no kriging estimates with which the results can be compared. Such will be the case in 'problem' mines, those in which the geology is complex,

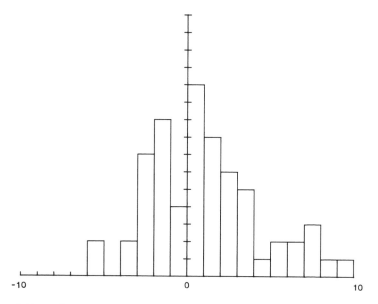

FIG. 5.18.   Histogram of deviations of the averaged weighted median block
estimates from the 'true' values for the 64 blocks.

the grades are patchy and very non-stationary and those in which
there were multiple phases of mineralisation. Although kriging
can be and has been used in such cases, there are many un-
resolved problems which could well be amenable to nonparametric geostatistics.

CHAPTER 6

Further Developments

A number of separate topics are discussed in this chapter; there are not necessarily any direct connections among them, but what they all have in common is that they comprise fairly simple extensions of the capabilities of nonparametric geostatistics; in addition, some of the ideas introduced may also be used with parametric geostatistics.

6.1 ALTERNATIVE DISTANCE MEASURES

Simple geometric anisotropy can be handled by nonparametric geostatistics without necessity for any alteration in the distance measure used. This is better than can be achieved with parametric geostatistics, in which a linear transformation of distance is required for such cases. More complex situations, however, require special treatment even in nonparametric geostatistics.

6.1.1 Folds and Pseudo-Folds
If a mineral deposit is stratigraphically controlled (e.g. coal, gold

reefs, or many iron ores) and the strata are folded, there is a readily identifiable geological structure which may be removed by mathematical modelling of an 'unfolding' process to restore the deposit to its inferred original geometry (Fig. 6.1) before continuing to model the grades or other quality parameters. Many studies have been carried out on models of mathematically continuous structural deformation, for example, by Ramsay (1967). Models such as these are of value (a) if the fold structure is itself sufficiently complex or intense as to warrant the effort of producing an 'unfolding' model, (b) if sufficient knowledge of the deformation history is available, and (c) if there was no remobilisation of the ore minerals during or after deformation to result in cross-cutting structures unrelated to the pre-folding structural surfaces (Fig. 6.2). Reasons (a) and (b) are self-explanatory. In case (c), if remobilisation has occurred, then the result is that there are parts of the deposit which cannot be 'unfolded' since they were never folded.

In other deposits, with the appearance of folds, curved tabular structures may result from a combination of other geological processes such as diffusion, flow of fluids in pressure or temperature gradients, and sedimentary or magmatic processes. Many porphyry copper deposits fall into this category, as do skarns around curved contact zones, and roll-front uranium deposits. In all such cases, there is no reason to consider attempting a mathematical 'unfolding' process, and yet there is a distinct anisotropy in the deposit which requires expression. A simple case is that of a porphyry copper deposit of paraboloid form (Fig. 6.3). For estimation purposes, Euclidean distances measured across the middle (such as line AB) have no meaning, since such distances cross two discontinuities as they pass out of and back into the ore zone. It is much better to measure distances along the paraboloid surface. Unfortunately, this is a non-trivial mathematical problem, currently under study by J-M Rendu in connection with the identical problem in parametric geostatistics (Rendu, pers. comm., 1980). It is probably possible in many cases to obtain an approximate solution.

If structural information on the upper or lower surface of the ore-body is available, then an interpolation method —parametric

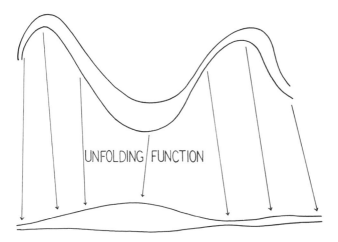

UNFOLDING FUNCTION

FIG. 6.1. Mathematical functions can be defined to transform a folded ore-body to an inferred unfolded geometry.

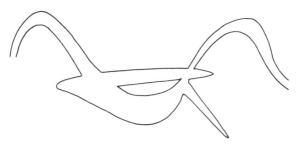

FIG. 6.2. Mathematical unfolding is not possible when cross-cutting ore bearing structures are present.

or nonparametric — may be used to obtain an estimated surface elevation for each block or cell. A numerical approximation to the length of a curved path in the surface between any two points could then be obtained by summation of the set of sloping line segments in all cells lying between the two points. This is, of course, not generally the shortest path, but to obtain that would

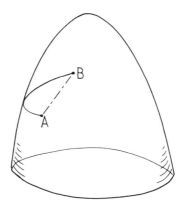

FIG. 6.3. A straight line is not the logical shortest path between two points (such as A and B) in a curved tabular ore-body such as a porphyry copper of paraboloid form.

require some global rather than merely local definition of the structure.

6.1.2 Discontinuities

It is not theoretically possible to handle discontinuities in either parametric or nonparametric geostatistics: there is no predictability and there should be no prediction across a discontinuity about which there is no information. However, in geology a large number of discontinuities are in fact faults which merely displace the rocks on either side; given the direction and amount of the displacement there is in theory no loss of information across the discontinuity. Therefore if the throw of a fault is known, *and* if the fault itself post-dates all of the relevant processes of ore genesis and migration, then the distances across the fault may be adjusted accordingly (by 'unfaulting', exactly analogous to 'unfolding' models mentioned in Section 6.1.1).

When the throw of a fault is unknown, or the discontinuity is not a fault, there is usually no alternative but to treat the distance between observations on opposite sides as infinite, and the predictability (reflected in the distance weighting) as zero.

6.2 MULTIPLE SURFACE INTERPOLATION

It is often required to generate models of a number of related surfaces, for example, the tops and bottoms of a set of coal seams within a stratigraphic unit. It is usual in such cases that the amount of information available varies from one surface to another, as illustrated in Fig. 6.4.

If two or more surfaces are, in fact, very closely related, then it might be possible in principle to combine the data for them simply by the addition or subtraction of constants. For instance, consider two coal seams separated by a known constant thickness of shale. The thickness of this interburden may be subtracted from each value recorded for the higher surface, the two sets of data may be combined and used to model the lower surface. However, the thickness is never constant enough for this procedure to be performed in practice.

One of the problems which is encountered in conventional single-surface modelling of multi-surface deposits is that because of the differences in data quality and quantity, there is a significant likelihood that modelled surfaces will cross, and that in a resource evaluation project there will be large errors due to

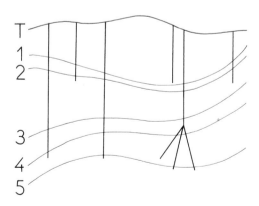

FIG. 6.4.   In a multi-seam coal deposit, the amount of data for the topography (T) and for each seam will be different.

omission of one of the seams in the 'crossover' area. It ought therefore to be possible to define a technique for computing a number
of surface models simultaneously in such a way as to avoid this
type of problem. Such a technique will be a constrained optimisa-
tion procedure in which the estimated value of the ith surface at
location P is defined as

$$\hat{x}_p(i) = f(u_p, \hat{x}_p(i-1), \hat{x}_p(i+1), x_1(i), \ldots, x_n(i))$$

in which the constraints might, for example, include

$$\hat{x}_p(i) - \hat{x}_p(i-1) \geq 0$$

$$\hat{x}_p(i+1) - \hat{x}_p(i) \geq 0$$

to prevent surface i from crossing either surface $i-1$ or $i+1$.
It would be quite easy to include in such a procedure any special
constraints which may be desired (such as $x_p(i) - x_p(i-1) = t$)
to force particular intervals to have constant thickness, or to
relax the constraints and allow particular seams to cross, if this is
deemed appropriate on the basis of the geological interpretation
(as it would be at a known unconformity). The procedure which
could be adopted would probably be a derivative of a non-linear
programming technique (a number of appropriate techniques are
discussed by Himmelblau, 1972); it would, however, be extremely
expensive to compute since all such methods are iterative, and
each iteration would require an ordinary geostatistical interpolation for each surface considered.

An alternative approach is more promising as a practical method.
As was pointed out in Chapter 4, the only assumption made about
the form of a surface for nonparametric geostatistics to be used
validly is that it be continuous. It need not have any explicit mathematical formulation. This means that substraction of any smooth
trend from the data before computation leaves a surface to be
estimated which is still continuous. Therefore, given two surfaces,
one known better than the other because it is defined by more data
points, the better known surface can be modelled using nonparametric geostatistical interpolation, and then it can be used as a
preliminary estimated trend to be subtracted from the less well

known surface which can be interpolated in its turn. If there are regions in the second surface which are better defined by data than the same regions of the first, then it is quite possible to repeat the process as many times as desired: estimating one surface and using it as a preliminary 'guessed' trend in fitting the other surface. The procedure is in fact not restricted to two surfaces, of course, and the nonparametric approach allows great flexibility in its use. It is even possible to add subjectively defined trends, though care must be taken to ensure that they relate to continuous surface models.

6.3 MULTIPLE VALUED SURFACES

Multiple valued surfaces lie outside the realms which are covered by parametric and nonparametric geostatistics so far, and perhaps will remain outside their scope. Yet the handling of such data is a significant problem in geology. Multi-valued surfaces are found for example in recumbent folds, in reverse faults and thrusts. If a standard nonparametric interpolation is attempted, on data drawn from a part of a surface which is indeed multi-valued, then the frequency distribution, at a point P, represented by the weighted observation values (if the values are of the surface elevation in the examples quoted above) will have more than one mode. Each mode will correspond approximately to the position of one of the leaves of the surface. If the nonparametric interpolation were computed, the result would be a single estimated surface, either a weighted median or a weighted adaptive quantile measure. If a parametric method were used, then again the interpolation would yield a single kriged estimate; these methods are based on the median and on the mean respectively. However, there is a third commonly used measure of the centre of a distribution — the mode. It is less useful statistically, and is difficult to define in mathematical terms, but the value corresponding to a peak on the probability density function (or histogram) is defined as a mode. If there are two or more peaks, then there are two or more mode values. One population can never, by definition of the terms, have more than one mean or median. Therefore the mode is an appropriate type of estimate to use in modelling multi-valued surfaces. Apart from this rôle, it may be that the mode has some other useful properties. One might be in the identification and location of discontinuities. Consider the calcium content of two different rock types — chalk and sandstone. Within geographic areas underlain by each rock type, there will be a small amount of variation in the calcium content. Across the discontinuity represented by the boundary between them, there is a very large change in calcium content. If chemical analyses were done on observations collected over the whole area irrespective of rock type, and an interpolation method invoked to produce a model of the calcium surface, the effect of either kriging or nonparametric geostatistical methods would be one of attempting to smooth values across the boundary. Using modes, the weighted frequency distribution close to the boundary on either side would be bimodal; the higher mode would reflect the values of the observations closest to the point to be interpolated; thus if only one mode (the higher) were taken, a 'weighted moving mode' method would generate a truly discontinuous surface, which would approximate the true surface very much better than either of the other techniques. Furthermore, like the nonparametric geostatistical techniques, it would be robust with respect to aberrant values. However, the mode is a very unstable measure with respect to the central part of the distribution, and has little in the way of 'nice' statistical properties to recommend it as a general purpose estimator. Moreover, since identification of a mode requires a reasonable well defined histogram, the data requirements of such a method are very much higher than for either parametric or nonparametric geostatistics.

CHAPTER 7

Summary

Within the general context of spatial data analysis, and particularly the problem of estimating point values and block
averages from regularly or irregularly scattered data, almost all
methods in routine use in the mining industry as in other fields
are derived from the ideas of standard parametric statistics. The
estimators that are used are all related in some way to the arithmetic mean, though the observations are usually weighted to
reflect a more or less objective measure of their relevance, and
the estimation may be carried out on data transformed in some
way to satisfy the distribution requirements of parametric
statistics. These ideas have been refined, to provide a set of
mathematically elegant solutions to the problem in a body of
theory called 'regionalised variable theory'. From this has come
a variety of related applications methods known as 'kriging'.
Together the theory and methods have been termed 'geostatistics'.
However, it has been shown in this volume that there is a quite
different possible approach to the same problems in spatial data
analysis, using some of the ideas of *non*parametric statistics, and
it has been suggested that a distinction should be drawn between
parametric and nonparametric geostatistics.

In studies where all the assumptions of parametric geostatistics are satisfied, regionalised variable theory guarantees optimal estimation, and nonparametric geostatistical methods are usually sub-optimal. However, in practice the assumptions are never fully satisfied and there are usually serious departures from them. In these cases, parametric theory can offer no assurances of the validity of its methods; nonparametric methods, however, continue to provide valid estimates provided only that the much less stringent assumptions hold true. There is, in fact, only one basic assumption in the nonparametric approach — that the spatial variation of the 'true value' is everywhere continuous — and even this is relaxed by the acceptance that point measurements of the true value may include errors. It is worthy of note that (in contrast with parametric statistics) the error distribution is not necessarily normal but may itself be any continuous distribution.

Starting from such simple assumptions, a nonparametric estimation problem would be approached as follows: for each point to be estimated, every available observation may be assigned a weighting which reflects an inferred probability that it is an independent member of a set of points defined as the local neighbourhood of the point to be estimated. There are two strands to the concept of 'independent membership': independence and membership. The probability of *membership* may be related purely to distance from the centre of the set, and some type of distance weighting can be adopted: it has been suggested in Chapter 4 that an inverse power weighting might be appropriate, with the power at least equal to the dimensionality of the observation space. It is quite possible, however, to use alternative distance weightings without changing the essence of the method. The probability of *independence* is related to the spatial clustering of the observations, and the probabilities of their belonging to each other's local membership sets; this is a function solely of the geometry of the sampling pattern, and may be computed once only for any set of observations. Because these two probabilities are in fact independent, the resultant weighting is merely their product. This estimate of independent membership of the neighbourhood set is the direct counterpart in nonparametric geostatistics to the kriging weight in parametric geostatistics.

Three different ways have been suggested for deriving point estimates from the set of weighted observations. The simplest uses the median of the distribution which they represent, computed as a weighted median of the observation values. It is known that this does not give an optimum estimate of the true value except when the true value is itself a monotonic function of the location. In a one-dimensional space this may be locally true, but in two- or three-dimensional space it is practically never the case. Because of this, some estimate is required of the proportions of the neighbourhood set which lie above and below the true value at the centre. This estimate may be obtained by a modification of Kendall's correlation coefficient, tau, to determine the relative proportion of ascending and descending relationships between pairs of observations nearer to and farther from the centre of the neighbourhood. The tau value resulting from this is used directly to define the quantile at which the true value is estimated to lie, again within the distribution represented by the set of weighted observations.

Although this 'weighted varying quantile' is theoretically an optimum nonparametric estimator, it suffers in fact from some instability in practice, particularly in zones close to individual observations, where a median would actually be a better and more stable estimator. For this reason, a hybrid method has been proposed, combining the stability of the median with the ability of the varying quantile to fit more closely the distribution of true values.

Although nonparametric geostatistics was developed intitally for point estimation from point observations, it is possible to extend it, by discretisation, to use extended observations (i.e. with finite spatial support), and to compute block average estimates. It may thus be expanded to fulfill the major functions of parametric geostatistics. The one field which is not catered for is that of 'structural analysis' in which a semivariogram is computed and its values plotted against distance of separation. The reason for this is that a variogram can only be computed if at least some form of stationarity is assumed. Since nonparametric geostatistics does not require *any* assumption of stationarity, there is no rôle in it for a variogram or for any nonparametric counterpart.

It may be inferred, from the foregoing discussion, that nonparametric geostatistics may be likened to a sledgehammer by comparison with the finely tuned precision instruments of parametric geostatistics. This may well be a true judgement, but a sledgehammer is indeed an appropriate tool for breaking real boulders.

References and Bibliography

Clark, I., 1979, *Practical Geostatistics*, Applied Science Publishers, London.

Clark, I. and Garnett, R. H. T., 1974, Identification of multiple mineralization phases by statistical methods, *Trans. Inst. Min. Metall.*, vol. 83, pp. A43–A52.

Cressie, N. and Hawkins, D. M., 1980, Robust estimation of the variogram, *Mathematical Geology*, vol. 12, no. 2, pp. 115–25.

Cubitt, J. M. and Henley, S., 1978, Statistical Analysis in Geology, *Benchmark Papers in Geology*, vol. 37, Dowden Hutchinson and Ross, Stoudsburg, Pa.

Cubitt, J. M. and Shaw, B., 1976, The geological implications of steady-state mechanisms in Catastrophe Theory, *Mathematical Geology*, vol. 8, no. 6, pp. 657–62.

Darling, D. A., 1957, The Kolmogorov–Smirnov, Cramer–von Mises tests, *Ann. Math. Stats*, vol. 28, pp. 823–38.

David, M., 1977, *Geostatistical Ore Reserve Estimation*, Elsevier, Amsterdam.

Davis, J. C., 1973, *Statistics and Data Analysis in Geology*, Wiley, New York.

Delfiner, P., 1976, Linear estimation of non-stationary spatial phenomena. In *Advanced Geostatistics in the Mining Industry*, (Guarascio, M., David, M. and Huijbregts, C., (Eds)), Reidel, Dordrecht, Netherlands, pp. 49–68.

Dougherty, E. L. and Smith, S. T., 1966, The use of linear programming to filter digitized map data, *Geophysics*, vol. 31, pp. 253–9.

Feller, W., 1957, *An Introduction to Probability Theory*, vol. 1, Wiley, New York.

Feller, W., 1966, *An Introduction to Probability Theory*, vol. 2, Wiley, New York.

Fieller, E. C., Hartley, H. O. and Pearson, E. S., 1957, Tests for rank correlation coefficients: 1, *Biometrika*, vol. 44, pp. 470–81.

Fisher, D. M., 1972, *Classification, Selection and Testing Procedures for Asymmetric Distributions*, unpublished PhD thesis, University of Iowa.

Gardiner, V. and Gardiner, G., 1978, Analysis of frequency distributions, *Concepts and Techniques in Modern Geography (CATMOG)*, No. 19, Geo Abstracts, University of East Anglia, Norwich.

Gibbons, J. D., 1971, *Nonparametric Statistical Inference*, McGraw-Hill, Kogakusha, Tokyo.

Guillaume, A., 1977, *Analyse des Variables Regionalisées*, Doin, Paris.

Harbaugh, J. W., Doveton, J. H. and Davis, J. C., 1977, *Probability Methods in Oil Exploration*, Wiley, New York.

Henley, S.,1976, Catastrophe Theory models in geology, *Mathematical Geology*, vol. 8, no. 6, pp. 649–55.

Himmelblau, D. M., 1972, *Applied Nonlinear Programming*, McGraw-Hill, New York.

Hogg, R. V., 1972, More light on the kurtosis and related statistics, *J. American Statistical Association*, vol. 67, pp. 422–4.

Hogg, R. V., 1974, Adaptive robust procedures: a partial review and some suggestions for future applications and theory, *J. American Statistical Association*, vol. 69, pp. 909–23.

Hogg, R. V., Fisher, D. M. and Randles, R. H., 1973, *A Two-Sample Adaptive Distribution-Free Test*, Technical Report 24, Dept. Statistics, University of Iowa.

Huber, P. J., 1972, Robust statistics: A review, *Ann. Math. Stats*, vol. 43, pp. 1041–67.

Journel, A. G. and Huijbregts, C. J., 1978, *Mining Geostatistics*, Academic Press, London.

Kendall, M. G.,1938, A new measure of rank correlation, *Biometrika*, vol. 30, pp. 81–93.

Krige, D. G., 1951, A statistical approach to some basic mine valuation problems on the Witwatersrand, *J. Chem. Metall. Min. Soc. South Africa*, vol. 52, pp. 119–39.

Krige, D. G., 1960, On the departure of ore value distributions from the lognormal model in South African gold mines, *J. South African Inst. Min. Metall.*, vol. 61, pp. 62–3, 231, 333.

Krige, D. G., 1978, Lognormal de Wijsian Geostatistics for Ore Evaluation, *South African Inst. Min. Metall. Monograph Series: Geostatistics*, vol. 1.

Lewis, P., 1977, *Maps and Statistics,* Halsted Press, New York.

Lindgren, B. W., 1976, *Statistical Theory* (3rd edn), Collier-Macmillan, London.

Link, R. F. and Koch, G. S. Jr., 1975, Some consequences of applying lognormal theory to pseudolognormal distributions, *Mathematical Geology*, vol. 7, no. 2, pp. 117–28.

Mann, C. J., 1970, Randomness in Nature, *Geol. Soc. Am. Bull.*, vol. 81, pp. 95–104.

Mann, H. B., 1945, Nonparametric tests against trend, *Econometrica*, vol. 13, pp. 245–59.

Matheron, G., 1962, *Traite de Geostatistique Appliquee*, tome 1, Editions Technip, Paris.

Matheron, G., 1963, Principles of geostatistics, *Economic Geology*, vol. 58, pp. 1246–66.

Matheron, G., 1973, The intrinsic random functions and their applications, *Advances in Applied Probability* , vol. 5, pp. 439–68.

Newton, R., 1973, A statistical prediction technique for deriving contour maps from geophysical data, *Mathematical Geology*, vol. 5, no. 2, pp. 179–89.

Olea, R. A., 1975, *Optimum Mapping Techniques Using Regionalized Variable Theory*, Series on Spatial Analysis, No. 2, Kansas Geological Survey, Lawrence, Kansas.

Olea, R. A., 1977, *Measuring Spatial Dependence with Semivariograms,* Series on Spatial Analysis, No. 3, Kansas Geological Survey, Lawrence, Kansas.

Ramsay, J. G., 1967, *Folding and Fracturing of Rocks*, McGraw-Hill, New York.

Randles, R. H. and Hogg, R. V., 1973, Adaptive distribution-free tests, *Communications in Statistics,* vol. 2, pp. 337–56.

Randles, R. H., Ramberg, J. S. and Hogg, R. V., 1973, An adaptive procedure for selecting the population with largest location parameter, *Technometrics*, vol. 15, pp. 769–78.

Rendu, J-M, 1978, An Introduction to Geostatistical Methods of Mineral Evaluation, *South African Inst. Min. Metall. Monograph Series: Geostatistics,* vol. 2.

Royle, A. G., 1978, Bias and its effects in ore-reserve estimators, *Trans. Inst. Min. Metall.*, vol. 87, pp. A8–A12.

Royle, A. G. and Newton, M. J., 1972, Mathematical models, sample sets, and ore reserve estimation, *Trans. Inst. Min. Metall.*, vol. 81, pp. A121–A128.

Royle, A. G., Newton, M. J. and Sarin, H. K., 1972, Geostatistical factors in design of mine sampling programmes, *Trans. Inst. Min. Metall.*, vol. 81, pp. A81–A88.

Sherman, B., 1950, A random variable related to the spacing of sample values, *Ann. Math. Stats.*, vol. 21, pp. 339–61.

Sichel, H. S., 1952, New methods in the statistical evaluation of mine sampling data, *Trans. Inst. Min. Metall.*, vol. 61, no. 6 pp. 261–8.

Siegel, S., 1956, *Nonparametric Statistics for the Behavioural Sciences,* McGraw-Hill, New York.

Snedecor, G. W. and Cochran, W. G., 1967, *Statistical Methods* (6th edn), Iowa State University Press, Ames, Iowa.

Thom, R., 1975, *Structural Stability and Morphogenesis,* (Fowler, D. H., (Trans.)), Benjamin, Reading, Mass.

Truscott, S. J., 1929, The computation of the probable value of ore reserves from assay results, *Trans. Inst. Min. Metall.*, vol. 38, pp. 482–96.

Tukey, J. W., 1962, The future of data analysis, *Ann. Math. Stats.*, vol. 33, pp. 1–67.

Tukey, J. W., 1970, Some further inputs. In *Geostatistics: A Colloquium,* (Merriam, D. F. (Ed.)), Plenum, New York.

Watermeyer, G. A., 1919, Applications of the theory of probability in the determination of ore reserves, *J. Chem. Metall. Min. Soc. South Africa,* vol. 19, pp. 97–107.

de Wijs, H. J., 1951, Statistics of ore distribution. Part 1: Frequency distribution of assay values, *Geologie en Mijnbouw,* vol. 13, pp. 365–75.

de Wijs, H. J., 1953, Statistics of ore distribution. Part 2: Theory of binomial distribution applied to sampling and engineering problems, *Geologie en Mijnbouw,* vol. 15, pp. 12–24.

de Wijs, H. J. 1973, Method of successive differences applied to mine sampling, *Trans. Inst. Min. Metall.*, vol. 82, pp. A78–A81.

Index